Advances in Polymer Science
Fortschritte der Hochpolymeren-Forschung

Volume 17

Edited by

H.-J. Cantow, Freiburg i. Br. · G. Dall'Asta, Cesano Maderno · J. D. Ferry, Madison · H. Fujita, Osaka · M. Gordon, Colchester · W. Kern, Mainz G. Natta, Milano · S. Okamura, Kyoto · C. G. Overberger, Ann Arbor G. V. Schulz, Mainz · W. P. Slichter, Murray Hill · A. J. Staverman, Leiden J. K. Stille, Iowa City

With 32 Figures

Springer-Verlag Berlin Heidelberg New York 1975

Editors

Prof. Dr. HANS-JOACHIM CANTOW, Institut für Makromolekulare Chemie der Universität, 7800 Freiburg i. Br., Stefan-Meier-Str. 31, BRD

Prof. Dr. GINO DALL'ASTA, SNIA VISCOSA—Centro Sperimentale, Cesano Maderno (MI), Italia.

Prof. Dr. JOHN D. FERRY, Department of Chemistry, The University of Wisconsin, Madison 6, Wisconsin 53706, U.S.A.

Prof. Dr. HIROSHI FUJITA, Osaka University, Department of Polymer Science, Toyonaka, Osaka, Japan

Prof. Dr. MANFRED GORDON, University of Essex, Department of Chemistry, Wivenhoe Park, Colchester C04 3SQ, England

Prof. Dr. WERNER KERN, Institut für Organische Chemie der Universität, 6500 Mainz, BRD

Prof. Dr. GIULIO NATTA, Istituto di Chimica Industriale del Politecnico, Milano, Italia

Prof. Dr. SEIZO OKAMURA, Department of Polymer Chemistry, Kyoto University, Kyoto, Japan

Prof. Dr. CHARLES G. OVERBERGER, The University of Michigan, Department of Chemistry, Ann Arbor, Michigan 48104, U.S.A.

Prof. Dr. GÜNTER VICTOR SCHULZ, Institut für Physikalische Chemie der Universität, 6500 Mainz, BRD

Dr. WILLIAM P. SLICHTER, Bell Telephone Laboratories Incorporated, Chemical Physics Research Department, Murray Hill, New Jersey 07971, U.S.A.

Prof. Dr. ALBERT JAN STAVERMAN, Chem. Laboratoria der Rijks-Universiteit, afd. Fysische Chemie I, Wassenaarseweg, Postbus 75, Leiden, Nederland

Prof. Dr. JOHN K. STILLE, University of Iowa, Department of Chemistry, Iowa City, U.S.A.

The format of the volumes in this series has been changed and slightly enlarged to allow a more rational use of paper. More text than before can now be printed on each page. We hope in this way to counter the effects of rising prices.

Das Format der Bände dieser Reihe ist geändert und geringfügig vergrößert worden. Das neue Format gestattet eine bessere Papiernutzung, da auf jeder Seite mehr Text untergebracht werden kann als zuvor. Wir wollen durch diese Rationalisierungsmaßnahme dem Preisanstieg entgegenwirken.

ISBN 3-540-07111-3 Springer-Verlag Berlin Heidelberg New York
ISBN 0-387-07111-3 Springer-Verlag New York Heidelberg Berlin

The use of general descrive names, trade marks, etc. in this publication, even if the former are not especially identified, is not to be taken as a sign that such names, as understood by the Trade Marks and Merchandise Marks Act, may accordingly be used freely by anyone.

This work is subject to copyright. All rights are reserved, whether the whole or part of the material is concerned, specifically those of translation, reprinting, re-use of illustrations, broadcasting, reproduction by photocopying, machine or similar means, and storage in data banks. Under § 54 of the German Copyright Law where copies are made for other than private use, a fee is payable to the publisher, the amount to the fee to be determined by agreement with the publisher. © by Springer-Verlag Berlin · Heidelberg 1975. Library of Congress Catalog Card Number 61-642. Printed in Germany. Typesetting and printing: Brühlsche Universitätsdruckerei, Gießen

Contents

Mechanical Synthesis of Block and Graft Copolymers
    ANTONIO CASALE and ROGER S. PORTER   1

Polymerization through the Carbon-Sulfur Double Bond
    WILLIAM H. SHARKEY   73

Mechanical Synthesis of Block and Graft Copolymers

ANTONIO CASALE

Montecatini-Edison SPA, Milano, Italy

ROGER S. PORTER

Polymer Science and Engineering
Materials Research Laboratory
University of Massachusetts, Amherst, Massachusetts 01002

Table of Contents

A. Introduction 2
B. Nomenclature 3
C. Mechanical Synthesis 4
D. Physical State in Mechanical Synthesis 7
   1. Solid State 7
      a) Polymer-polymer Systems 7
      b) Polymer-monomer Systems 8
      c) General Features of Vibromilling 8
         Natural Polymers 8
         Vinyl Polymers 9
         Polyacetals 14
         Polyamides 15
         Polyesters 19
         Polycondensation 24
   2. Rubbery State 30
      a) Polymer-polymer Systems 30
         Elastomers 30
         Plastomers 33
      b) Polymer-monomer Systems 35
      c) Plastomers and Natural Polymers 48
   3. Molten State 59
   4. Solution Systems 62
      a) High Speed Stirring 62
      b) Freezing and Thawing 64
      c) Vapor Phase Swelling 64
      d) Spark Discharge 66
   5. Polymer-Filler Interactions 67
E. References 69

The authors express their appreciation to the U.S. Army Research Office — Durham, and to Montecatini-Edison.

Where illustrations are taken from other publications, the source is indicated by the number of the literature reference.

A. Introduction

Over the last two decades, demands have increased dramatically for the development of polymers with an increasing variety of complex compositions. The goal has been to develop polymers for new areas of application with the limitation of a restricted number of monomers due to cost. Polymer scientists have thus initiated studies with the purpose of changing polymer properties by combining them in a variety of ways. Such combinations of polymers may be either true solutions or heterophase systems. A large body of literature covers these systems. The total range of polymer properties, however, is beyond the limits of polymer blends with their limitations of mutual solubility and compatibility. An interesting possibility for overcoming this problem is thus afforded by polymer modification including changes in chemical composition and structure. This can be achieved by block and graft polymerization. These techniques permit a combination of different properties in one material; *e.g.*, heat resistance and elasticity; hydrophobic and hydrophilic properties; and combinations of crystalline and amorphous structures.

Some of the major areas covered by the block and graft copolymers involve the possibility of:
(1) modifying polymers such as polystyrene, poly(styrene-coacrylonitrile), poly(vinyl chloride), poly(methyl methacrylate) by adding small amounts of elastomer to improve toughness;
(2) reinforcement of rubbers by grafting resins with different active groups or fillers.

Such copolymers may also be effectively employed to enhance adhesion between surfaces of different composition using copolymers with groups having different activity. These complex systems obtained by reaction (graft, block, homopolymers) have been termed interpolymer. For these and additional reasons, increasing amounts of block and graft copolymers are produced by the major plastics companies throughout the world.

Various methods of graft and block copolymerization have been described in the literature (radical, irradiation, condensation, etc.). The reader is referred to the specialized books (*1–3*) in regard to synthesis, characterization and properties.

Of all the methods for the production of block and graft polymers, the greatest importance, from the view of commercial simplicity, involves mechanical synthesis. The block and graft reactions can be potentially performed directly during polymer processing and in standard equipments, such as internal mixers, injection molding machines, and extruders.

Another advantage of mechanical synthesis is the minor influence of production cost. This allows the tailoring of polymers for specific needs of industry where small amounts of interpolymers of different composition are required.

The synthesis of block and graft copolymers by mechanical forces has been earlier reviewed by many authors (*4–10*). This review is thus not intended to be exhaustive and a different approach has thus been taken. *The material is organized by state of matter rather than by the instruments used.* Thus the authors hope to provide a descriptive mosaic for the parallel studies developing in different countries and in different facets of the field. Some of the original references may be difficult to procure so that more general secondary references are cited in some cases.

Intercomparisons are inevitably qualitative since mechanically induced reactions have been performed in widely divergent instruments which, in some cases, reveal woeful uncertainties in shear mode, history and intensity. Some of the methods are, however, amenable to commercial scale-up such as coextrusions in the rubbers or molten states, whereas others, for example ball milling provide more information about the potential of preparing grafts and blocks. We recognize that there are other meanings for the word mechanochemistry. It is, however, widely used in this field. By using a distinctive title and outline, we expect that readers will not be readily misled.

B. Nomenclature

The nomenclature system used here is that suggested by Ceresa (*1*) and adopted in the Encyclopedia of Polymer Science and Technology. A block copolymer can be represented by:

~ AAAAAAAAAAAAAAAABBBBBBBBBBBBBBBBBBAAAAAAAA ~

and a graft copolymer by:

```
~ AAAAAAAAAAAAAAAAAAAAAAAAAAAAAA ~
      B                      B
      B                      B
      B                      B
      B                      B
      B                      B
      B
```

The sequence of A units in the graft copolymers is referred to as the backbone and the branches of B units as the grafts. In more sophisticated cases, the backbone and/or the side chain can themselves be copolymers. The reader is referred to the Encyclopedia for details.

In identifying chain segments, the term -co- is interposed between the names of the monomers copolymerized; the letter -b- is used to designate block copolymers. Similarly the letter -g- is used to indicate grafted segments. The first polymer named corresponds to the polymer prepared in the first stage of synthesis.

C. Mechanical Synthesis

The mechanical synthesis of block and graft copolymer is a method of sizable versatility. It can be performed (as already stated) during polymer processing and in standard equipment. The reaction, also, can be carried out by subjecting a mixture of two or more polymers to mechanical degradation in either the solid, elastic-melt, or solution states. It is, also, possible to induce reaction mechanically between polymers and monomers.

The characteristic feature of these reactions is the direct conversion of the mechanical work, A, used in deforming the macromolecules, into the chemical energy of the activated chains and radicals formed.

$$A = (\partial_2 - \partial_1) \Delta n$$

where ∂_1 and ∂_2 are the chemical potentials of the system before and after mechanical breakdown, Δn is the increase in the number of active macromolecules in the mechanical reaction (5). Lowering of the energy barrier is then achieved through mechanical work, and the reaction does not require supplemental energy (*e.g.*, heat and light).

All of the syntheses methods described in earlier reviews, (comminution, vibromilling, mastication, extrusion, stirring, ultrasonic irradiation, freezing and thawing, discharging of high voltage sparks, and swelling) (*11–13*) have been employed to synthesize block and graft polymers, using, as initiator, the polymer radical formed by the mechanical scission of macromolecules. Recently, the possibility has also been demonstrated for polycondensation reactions induced mechanochemical activation. This reaction will be discussed later in detail.

The structure of copolymers obtained by mechanical synthesis depends on the properties of the components at reaction conditions. By altering conditions, it is possible to regulate, within wide limits, the rate of polymer breakdown and thus influence the composition of resulting products.

It is generally believed that the mechanical synthesis involves the following general steps with the composition of the resulting interpolymers depending on the relative rates:

1. Mechanical scission

$$P_m - P_n \rightarrow P_m^{\cdot} + P_n^{\cdot}$$

$$R_s - R_t \rightarrow R_s^{\cdot} + R_t^{\cdot}.$$

2. Recombination

$$P_m^{\cdot} + P_n^{\cdot} \rightarrow P_m - P_n$$

$$R_s^{\cdot} + R_t^{\cdot} \rightarrow R_s - R_t.$$

3. Cross Combination

$$P_n^{\cdot} + R_s^{\cdot} \rightarrow P_n - R_s \quad P_n^{\cdot} + R_t^{\cdot} \rightarrow P_n R_t$$

$$P_m^{\cdot} + R_t^{\cdot} \rightarrow P_m - R_t \quad P_m^{\cdot} + R_s^{\cdot} \rightarrow P_m R_s.$$

4. Scission of chains by mechanically activated macroradicals

$$P_n^{\cdot} + R_s - R_t \rightarrow P_n - R_s + R_t^{\cdot}.$$

5. Chain transfer from polymer radicals to macromolecules

$$P_n^{\cdot} + R - R - R \rightarrow P_n + R - R^{\cdot} - R.$$

6. Termination by disproportionation

$$P_n - R_s^{\cdot} + P_m - R_t^{\cdot} \rightarrow P_n - R_s + P_m - R_t.$$

7. Termination by reaction with solvent, radical acceptors, oxygen and so on.

In the absence of substantial unsaturation and of active groups on the chain for either polymer, only linear block copolymers are formed, according to the initiation Reactions 1 and 4. Low density polyethylene and high molecular weight polyisobutylene are typical of polymers which form block copolymer fractions on intensive mechanical working. The composition of block copolymers is related also to the relative rates of reaction, (Reactions 2 and 3) which is determined by the relative radical reactivity.

According to Berlin (5), Reaction 4 can predominate for the following reasons:
(1) The high yield of block copolymers cannot be explained by the relatively low concentration of active molecules formed by mechanical degradation.
(2) The probability of a reaction between the small number of macroradicals and the large quantity of macromolecules is high.
(3) The activation of the macromolecules as a consequence of valence angle deformation by mechanical action.

(4) The initiation rate is proportional to the concentration of macroradicals and of the macromolecules. In view of this, the initial scission reactions may proceed fairly intensively even with low macroradicals concentration, as the macromolecules concentration is high. As a result of the initiating scission, block copolymers are formed and new macroradicals are created which continue this chain process further.

A block copolymer may initially have the simplest structure $P-R$, but, as the molecular weight of the homopolymers falls during mastication, multisegment block copolymers are formed, by mechanical scission of initial reaction products.

If one of the polymer chains possesses active groups (*e.g.* double bonds, halogen atoms, and methylene groups) graft copolymerization of the ruptured chain can occur randomly along the chain of the second polymer according to Reaction 5. If both of the polymers can undergo rupture and possess active groups very complex structures are formed. If one of the polymers gels when masticated, a crosslinked gelled structure can be achieved to which there are grafted branches of the second component (*11*).

If the mechanochemical reaction is performed in the presence of monomers, the polymeric radicals can initiate the polymerization of the monomer present:

$$P_n^{\cdot} + sR \rightarrow P_n - R_s^{\cdot}.$$

The progressive decrease of the initial polymer molecular weight leads to segments of block copolymer, and to homopolymer of the monomer:

$$P_n - R_s \rightarrow P_n - R_{s-x}^{\cdot} + Rx^{\cdot}$$

or

$$P_n - R_s \rightarrow P_{n-x} + P_x - R_s.$$

As a consequence, multisegment block copolymers may be formed.

The higher the initial polymer molecular weight and the monomer tendency to terminate by combination, the more complex the block copolymer structure.

If the macroradicals are stabilized by transfer to monomer, the tendency of the monomer to homopolymerize or to block copolymerize depends on the reactivity ratio of the monomer towards the two radicals.

As is well known from free radical copolymerization theory, the composition of the copolymers will depend only on the propagation reaction. The relative ability of monomer to add to a growing chain is influenced by the nature of the last chain unit and by the relative concentration. Generally, chain transfer to monomer by polymer radicals will occur to an appreciable extent, and the final product will be made up of homopolymers, multisegment block copolymers, and branched and grafted structures. In the presence of two or more monomers, the copolymer composition will depend on the relative monomer-radical reactivity ratios and concentration.

If the degradation reaction is carried out in the presence of an unpolymerizable species, only one molecule of the monomer adds to the primary radical. This is the case for the system natural rubber-maleic anhydride (*11, 65, 71*).

The picture of mechanical synthesis is much more complex if the segments of block and graft copolymers can undergo rupture to polymeric free radicals forming multisegment block copolymers, gelled and crosslinked structures. Baramboim in his book (*11*) describes 14 different possibilities of block and graft reactions.

D. Physical State in Mechanical Synthesis

1. Solid State

The method of obtaining graft and block copolymers by grinding or comminution has the advantages of permitting an unlimited combination of different polymers (as well as monomers) and of effecting reactions without solvents. Limitations of the mechanical process are a relatively high power consumption, equipment complexity and metal abrasion. The abraded metal is generally not inert to macroradicals and in some cases chemically "grafts" in itself, modifying the polymer properties. Thus, the practical development of this method depends on the design of efficient apparatus for mechanical comminution (*4*). Practically speaking, the only equipment used is the vibromill (*11–13*). In a very few cases a standard ball mill has been employed.

a) Polymer-polymer Systems

The vibromilling of blends of two or more polymers yields block and graft copolymers when the polymeric radicals, formed mechanically react by combination rather than by disproportionation. However, only a few such experiments have been run.

Deters (*14*) vibromilled a blend of cellulose and cellulose triacetate. The acetic acid content of cellulose acetate decreased with grinding time (40 h) while that of the cellulose increased, suggesting the formation of a block or graft copolymer or of an esterification reaction by acetic acid developed by mechanical reaction. Baramboim (*15*) dissolved separately in CCl_4 polystyrene, poly(methyl methacrylate), and poly(vinyl acetate). After mixing equal volumes of solutions of equivalent polymer concentration, the solvent was evaporated at 50° C under vacuum and the resultant product ball-milled. The examination of the ball-milled products showed the formation of free radicals which copolymerized.

Bischof (*16*) used the macroradicals resulting from vibromilling as initiators for synthesis of block and graft copolymers of poly(methyl methacrylate) with poly(vinyl chloride) with polyacrylonitrile.

b) Polymer-monomer Systems

The mechanical synthesis of block and graft copolymers by vibromilling a polymer-monomer blend has been performed by many researchers. Natural polymers (*14, 17*) vinyl polymers (*18–27*), and heterochain polymers (*18, 28–34*) have been formed during polymer mechanochemical degradation. Importantly, Simionescu, Vasiliu-Oprea and Neguleanu studied the possibility of carrying out mechanically-induced polycondensations starting from polyesters and diamines (*33, 35–37*).

c) General Features of Vibromilling

Monomers may be in the solid, liquid or gaseous state. In the first phase, blending and comminution is generally carried out to reduce inhomogeneities. For the same reason, polymer particle size can also influence the reaction. For example, in the grafting of vinyl chloride on polycaprolactam, the Cl content of the resultant polymer was 3.71% and 2.16% when 0.05–0.09 mm and 0.4–0.63 mm diameter polycaprolactam particles were employed (*31*). To obtain polymer in the desired powder form, the polymer was pulverized and selectively precipitated.

Many reactions have been performed in the presence of a solvent. However, the solvent must be chosen carefully to avoid reaction with polymer. For example, the low yield for grafts of acrylonitrile on polyamides in the presence of methanol has been shown to be due to the methanolysis (*18, 31*). Generally speaking, the grafted products are principally obtained; however minor amounts and homopolymers can also result. The homopolymerization proceeds by an intramolecular transfer reaction between macroradicals and monomers. The amount of homopolymer depends on the system. Details on systems already investigated will be described in the next section.

Mechanical syntheses are, of course, generally affected by the presence of radical acceptors (*20, 29*). The yield on copolymers also increases with duration of mechanical stress. However, if the milling time is too long, the properties of the graft or block copolymers can be deteriorated by degradation of initial product. The nature of the balls and mill material can also influence the reaction (*37*); the mechanical activation of inorganic materials. The production of graft and block copolymers on freshly formed surfaces has been established in the lierature.

Natural Polymers. Whistler and Goatley (*17*) investigated the possibility of using free radicals generated by ball milling of corn starch as initiator for acrylamide polymerization. They used a porcelain mill at 55 rpm for 150 h at 26° C in an inert atmosphere. In the case of corn amylose the ratio polymer/monomer/solvent (dry methanol) was 1/0.5/0.3; in the case of waxy corn starch, 1/0.3/0.5. In this last case the solvent was ethanol. A Tiselius electrophoresis pattern on grafted whole corn starch contained two peaks, suggesting grafts of amylose and amylopectine, while a pattern from a waxy corn starch graft showed a single peak as expected for grafted amylopectine.

Deters (14) grafted acrylonitrile, methyl methacrylate and vinyl chloride on cellulose and cellulose triacetate. The first two monomers were put in the reactor as liquids, the last as a gas. The results are summarized on Table 1. Vinyl chloride did not graft to cellulose (14).

Table 1. Polymerization of vinyl polymers by vibromilling cellulose derivatives. Composition and properties of the interpolymers (14)

Polymer	Monomer	Product fraction		Content of respectively N, acetic acid, or Cl, %	Graft polymer %
		Type	%		
Cellulose	Acrylonitrile	Cel–AN I	47.0	23.1	89.5
		Cel–AN II	26.4	12.2	47.0
		Cel–AN III	26.4	1.3	5.0
Cellulose triacetate	Acrylonitrile	CTA–AN I	62.5	18.6	72.0
		CTA–AN II	9.4	16.9	62.5
		CTA–AN III	28.0	1.1	4.3
	Methyl methacrylate	CTA–MMA I	—	0.4	99.4
		CTA–MMA Ia	—	3.0	95.3
		CTA–MMA II	—	51.4	17.7
	Vinyl chloride	CTA–CV I	59.8	15.8	27.7
		CTA–CV II	40.2	46.9	83.0

Vinyl Polymers. Vibromilling of a vinyl polymer in the presence of a polymerizable monomer can yield both graft and block copolymers. A typical example is the graft synthesis caused by vibromilling poly(methyl methacrylate) in the presence of gaseous vinyl chloride at 25° C for 12 h (19–21, 38).

The polymeric radicals resulting from the mechanical scission of poly(methyl methacrylate) are:

$$\sim CH_2-\underset{\underset{O}{\overset{\|}{COCH_3}}}{\overset{CH_3}{C}}-CH_2-\underset{\underset{O}{\overset{\|}{COCH_3}}}{\overset{CH_3}{C}}-CH_2 \sim \longrightarrow \sim CH_2-\underset{\underset{O}{\overset{\|}{COCH_3}}}{\overset{CH_3}{\overset{|}{C}}}\cdot \quad + \quad \sim CH_2-\underset{\underset{O}{\overset{\|}{COCH_3}}}{\overset{CH_3}{\overset{|}{C}}}-CH_2 \sim$$

$$(1) \qquad\qquad (2)$$

The less stable radical (2) may disproportionate [according to Todd (39)] to:

$$\sim CH_2-\overset{CH_3}{\underset{}{C}}=CH_2 + \cdot COOCH_3 \quad \text{or} \quad \sim CH_2-\underset{COOCH_3}{\overset{}{C}}=CH_2 + \cdot CH_3$$

The following reactions may occur: combination, recombination, reaction with radical acceptors, disproportionation and transfer reactions.

In the presence of monomers, of course, graft and block copolymers are formed. The polymerization is initiated by the macro free radicals generated by mechanical stresses (block copolymers) or by free radicals obtained by intermolecular transfer (graft polymers), such as

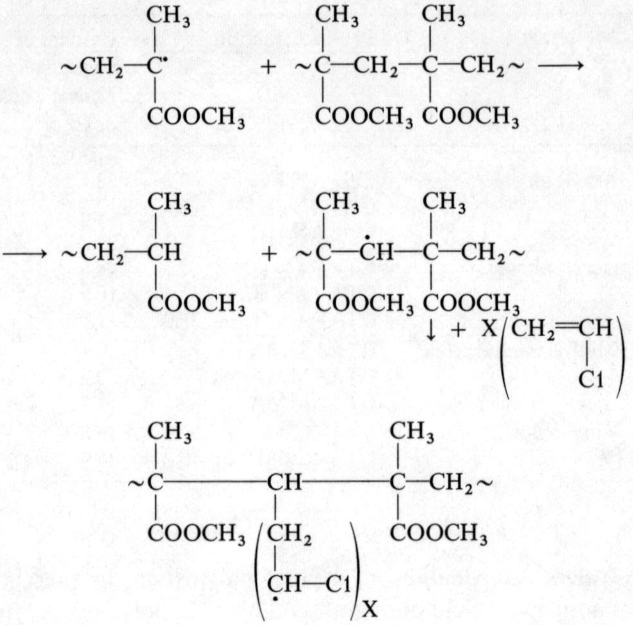

$$\sim CH_2-\underset{\underset{COOCH_3}{|}}{\overset{\overset{CH_3}{|}}{C^\bullet}} + \sim\underset{\underset{COOCH_3}{|}}{\overset{\overset{CH_3}{|}}{C}}-CH_2-\underset{\underset{COOCH_3}{|}}{\overset{\overset{CH_3}{|}}{C}}-CH_2\sim \longrightarrow$$

$$\longrightarrow \sim CH_2-\underset{\underset{COOCH_3}{|}}{\overset{\overset{CH_3}{|}}{CH}} + \sim\underset{\underset{COOCH_3}{|}}{\overset{\overset{CH_3}{|}}{C}}-\overset{\bullet}{CH}-\underset{\underset{COOCH_3}{|}}{\overset{\overset{CH_3}{|}}{C}}-CH_2\sim$$

$$\downarrow + X\begin{pmatrix} CH_2{=}CH \\ | \\ Cl \end{pmatrix}$$

$$\sim\underset{\underset{COOCH_3}{|}}{\overset{\overset{CH_3}{|}}{C}}-------CH-------\underset{\underset{COOCH_3}{|}}{\overset{\overset{CH_3}{|}}{C}}-CH_2\sim$$

$$\begin{pmatrix} CH_2 \\ | \\ CH-Cl \end{pmatrix}_X$$

However, if increasing amounts of free radical acceptors are added to the system, the block and graft polymerization gradually decreases, as saturation of free macroradicals proceeds faster than the addition of monomers. In Table 2, the influence of benzoquinone on the graft polymerization of vinyl chloride on poly(methyl methacrylate) is shown (20).

Table 2. Polymerization of vinyl chloride by poly(methyl methacrylate) vibromilling. Effect of benzoquinone on the Cl content of the interpolymer (20)

Benzoquinone amount, mg	Benzoquinone % of poly(methyl methacrylate)	Cl content %
0.0	0.000	18.6
10.0	0.033	14.2
50.0	0.166	6.85
1000.0	3.330	0.6

The reactions are:

1. $-CH_2-\underset{\underset{COOCH_3}{|}}{\overset{\overset{CH_3}{|}}{C^{\bullet}}} + O=\bigcirc=O \xrightarrow{k_1} -CH_2-\underset{\underset{COOCH_3}{|}}{\overset{\overset{CH_3}{|}}{C}}-O-\bigcirc-O^{\bullet}$

2. $-CH_2-\underset{\underset{COOCH_3}{|}}{\overset{\overset{CH_3}{|}}{C}}-O-\bigcirc-O^{\bullet} + {}^{\bullet}\underset{\underset{COOCH_3}{|}}{\overset{\overset{CH_3}{|}}{C}}-CH_2-$

$\downarrow k_2$

$-CH_2-\underset{\underset{COOCH_3}{|}}{\overset{\overset{CH_3}{|}}{C}}-O-\bigcirc-O-\underset{\underset{COOCH_3}{|}}{\overset{\overset{CH_3}{|}}{C}}-CH_2-$

where $k_1 > k_2$

Figure 1 shows the Cl content of the interpolymer as a function of milling time and temperature. The vibromilling technique was also applied to the system poly(methyl methacrylate)-acrylonitrile at 40° C. The results are shown in Table 3 (*19*).

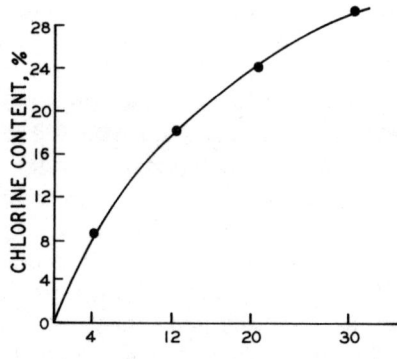

Fig. 1. Polymerization of vinyl chloride by vibromilling poly(methylmethacrylate). Effect of milling time in hours on Cl content (*20*)

Table 3. Polymerization of acrylonitrile by vibromilling poly(methyl methacrylate). Composition of the interpolymer fractions (19)

Time h	Composition %		Benzene insoluble fraction composition %			Benzene soluble fraction composition %		
	PMMA	PACN	%	PMMA	PACN	%	PMMA	PACN
8	85.2	14.8	85.5	80.2	19.8	14.5	97.3	2.7
16	81.1	18.9	74.75	79.0	21.0	25.25	97.6	2.4
24	79.6	20.4	64.9	60.3	29.7	35.1	95.8	4.2

PMMA = poly(methyl methacrylate).
PACN = polyacrylonitrile.

Poly(methyl methacrylate) was also subjected to mechanical reaction in a vibrating mill in common solvent for several monomers (ethylene, acrylic acid and its esters, acrylonitrile and styrene) at temperatures from -196 to $20°$ C (22). The formation of macroradicals and their reactions were followed by EPR (electron paramagnetic resonance). The macroradicals reacted with vinyl monomers at temperatures less than $-100°$ C, while quinones underwent reaction as low as $-196°$ C. The same experiments were performed also with polystyrene and polybutylenedimethacrylate. The radicals from polystyrene were more reactive than those from poly(methyl methacrylate).

The grinding of frozen solutions of poly(methyl methacrylate) in acrylic acid gave a hygroscopic grafted polymer (23). About 80% of poly(methyl methacrylate) was grafted. The structure of the polymer was determined by acid hydrolysis and potentiometric titration. The graft polymers after hardening with hexamethylene diamine gave hard films which exhibited a good adhesion to glass and steel. The same technique was applied to a dispersion of 30% poly(vinyl alcohol) in acrylic acid (24).

The ball milling of poly(vinyl chloride) and acrolein gave a product with improved heat stability and electrical properties (25). The milling was performed at 108 rpm and at room temperature for four hours using porcelain balls 1.2–2 cm in diameter. After milling with 20% of acrolein for two hours, the decomposition temperature of the polymer product increased from 150 to 160° C and the darkening temperature from 200 to 220° C. After four hours the two temperatures were, respectively, 220 and 280° C. The ball milled product, also, showed a 37-fold improvement in volumetric resistance and a 15 000-fold improvement in surface electrical resistance over the original poly(vinyl chloride).

Polyacrylonitrile was ground under a constant pressure of one atm with vinyl chloride and with butadiene to give graft and block copolymers as well as minor amounts of homopolymer in the first system (27). The products were characterized by chemical and infrared analysis, viscometric and turbidimetric measurements, and solubility. The results are reported in Table 4.

A detailed study on the possibility of modifying polyethylene by mechanical method was performed by Protasov and Baramboim (26). In their investigation they used a low density polyethylene of M_v 35400 and several solid monomers

Table 4. Polymerization of various monomers by acrylonitrile vibromilling (27)

Time h	Vinylchloride amount on interpolymer	Butadiene amount on interpolymer
2	1.1	—
6	6.6	—
7	—	9.0
10	7.6	—
20	14.0	11.2
30	24.0	18.0
60	25.5	—
180	32.7	—

Table 5. Polymerization of solid monomers by vibromilling polyethylene. Composition and properties (26)

Monomer	Milling time min	Yield of copolymer	Character of copolymer	Copolymer (comp.) % PE	Modifying monomer	Total exchange capacity mg equiv./g
Acrylic acid	5	48.0	Crosslinked	60	40	5.25
Acrylonitrile	3	12.3	Crosslinked	—	—	—
Methacrylamide	6	13.3	Crosslinked	51	49	1.46
Acenaphtylene	5	—	Linear	—	—	6.46[a]
Maleic anhydride	5	0.3[b]	Linear	7	3	—

[a] After sulphonation.
[b] Amount of graft.

AA acrylic acid
AN acrylonitrile
MAA methacrylamide
AC acenaphtylene
MA maleic anhydride

(methacrylamide, maleic anhydride, acenaphthylene, acrylic acid and acrylonitrile). The mechanical dispersion was performed in a four-position eccentric vibromill with cavities and balls, diameter 6–8 mm, made of the same stainless steel. The vibration frequency was 50 Hz with an amplitude of 2–4 mm. The reaction was carried out in air at temperatures from $-30°$ to $-65°$ C. The formation of copolymeric products, after separation by extraction, was confirmed by many techniques (qualitative reaction of functional groups, elementary analysis, acid number, nitration, potentiometric titration and IR spectroscopy).

The copolymerization of polyethylene under this condition can be considered a variation of solid-phase mechanochemical synthesis. Table 5 gives some results.

The following points can be underlined:
1. The reaction rate is high, in spite of system heterogeneity.
2. After monomer conversion, apparently further formation of copolymers may be effected by a mechanism of mechanosynthesis in the polymer-polymer system.

3. The susceptibility order of monomers to mechanosynthesis (acrylic acid > acrylonitrile > methacrylamide > acenaphthylene > maleic anhydride) is consistent with related chemistry. Maleic anhydride, which has the lowest susceptibility and is not homopolymerizable, probably combines as a single monomer unit.
4. The products of polyethylene modification are mixed block and graft copolymers of three dimensional structures (with acrylic acid, methacrylamide and acrylonitrile) or linear structures (acenaphthylene and maleic anhydride). This is again consistent with the nature of the monomers and the activity of their radicals.
5. Polyethylene properties can be markedly changed and in accordance with the coreacted monomer; particularly affected are the hydrophobic and thermomechanical properties.

Polyacetals. Simionescu and coworkers (28) have extended to polyoxymethylene the process of grafting vinyl polymers (acrylonitrile and methylmethacrylate). They performed the synthesis using a virbomill at room temperature under vacuum (10.1 Torr). The initial monomer-polymer ratio was 1.5/1.0 and the degree of vibromill packing 0.44. Before milling the polyoxymethylene granules were dissolved in dimethyl formamide and reprecipitated with the aim of stabilizer removal and for reduction of polymer particle size (from 2–2.5 mm to 0.05–0.10 mm). Full details of the reaction have been described (40).

Figure 2 shows the acrylonitrile conversion to homopolymers and the nitrogen content of the graft polymers as a function of time (28). The formation of the graft polymers was determined by IR analysis. The molecular weight of the grafted chain was determined by intrinsic viscosities on pyrolyzed (200° C) graft polymers. The acrylonitrile segments survive pyrolysis and had a molecular weight of about 2000 after 48 h of milling. The mechanical reaction changed profoundly the properties of polyoxymethylene especially as regards to heat stability and acid, base and phenol solubility. Figure 3 shows the change of heat resistance as a function of the amount of polyacrylonitrile grafted. The grafted polymer maintained a high degree of crystallinity (28).

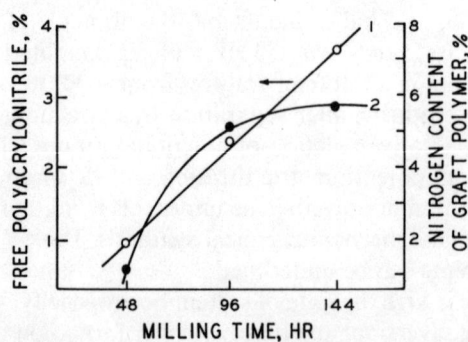

Fig. 2. Polymerization of acrylonitrile by vibromilling polyoxymethylene. Effect of milling time in hours on *1*) free polyacrylonitrile; *2*) nitrogen content of graft copolymer (28)

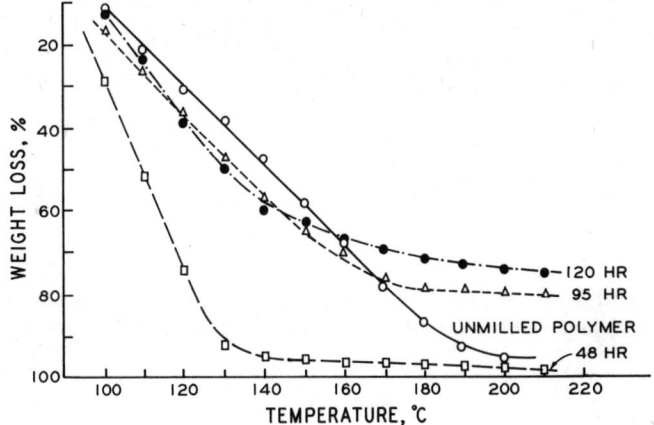

Fig. 3. Polymerization of acrylonitrile by vibromilling polyoxymethylene. Thermogravimetric analysis of the polyoxymethylene (without stabilizers) and of the interpolymers at different milling times (28)

The grafting of methylmethacrylate was studied, under the same condition as acrylonitrile (28). After 24 h of milling there was an increase in methacrylate homopolymer (see Fig. 4). This is probably due to the mechanochemical degradation of the grafted chains in the last stage of reaction. The heat stability of the graft polymer is better than that of the original polymer or of the same polyoxymethylene grafted with acrylonitrile.

Polyamides. As already described, polyamide degradation takes place in the dry state at the C—C bonds, whereas in the wet state an activated mechanochemical hydrolysis predominates. However, only the macroradicals obtained from —C—C— bonds are able to initiate graft reactions.

$$NH_2- \ldots -CH_2^{\cdot} + CH_2\!\!=\!\!CH \longrightarrow NH_2- \ldots CH_2-CH_2-CH^{\cdot}$$
$$\qquad\qquad\qquad\qquad\quad | \qquad\qquad\qquad\qquad\qquad\qquad |$$
$$\qquad\qquad\qquad\qquad\quad R \qquad\qquad\qquad\qquad\qquad\qquad R$$

$$NH_2- \ldots CH_2-CH_2-CH^{\cdot} \xrightarrow{monomer} NH_2- \ldots CH_2\text{-}[CH_2-CH]_n CH_2-CH^{\cdot}$$
$$\qquad\qquad\qquad\qquad | \qquad\qquad\qquad\qquad\qquad\qquad\qquad\quad | \qquad\quad |$$
$$\qquad\qquad\qquad\qquad R \qquad\qquad\qquad\qquad\qquad\qquad\qquad\quad R \qquad\quad R$$

Baramboim and coworkers (29, 30) studied the possibility of grafting solid acrylic acid on polyamide, AK 60/40 by vibromilling at low temperature. Polyamide was dissolved in acrylic acid, then the 25% solution was frozen at dry ice temperature and ground for three minutes at 50 cycles/second. The reaction was completed in one minute and the mechanochemical degradation of the grafted polymer subsequently occurred. The structure of the polymers was studied by

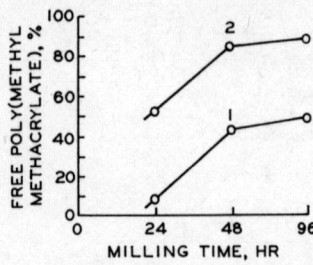

Fig. 4. Variation of homopolymer (PMMA) amount in the product after grafting (Curve *1*), and conversion of monomer (MAM) to homopolymer at different milling times (Curve *2*) (*28*)

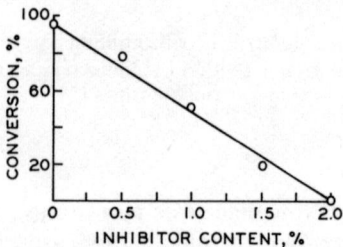

Fig. 5. Polymerization of acrylic acid by vibromilling a frozen solution of polyamide. Effect of inhibitor content on conversion (*29*)

extraction with solvents, chemical analysis, and by acid hydrolysis of the heterochains. The mixture was composed of 30% water soluble polymer with a nitrogen content of about 1.2% and of 70% of water insoluble block and graft copolymer. No polyamide was extracted. Though the copolymer forms an opaque aqueous gel, it does not have a crosslinked structure, as it is completely soluble in solvents. Analysis showed that the composition of the whole copolymeric product is close to that of the original frozen solution.

The molecular weight of the grafted chain of polyacrylic acid, determined by acid hydrolysis, is of the order of 10000. The product differs markedly from the original polyamide in electrochemical properties. As a confirmation of a radical mechanism, radical acceptors prevented the reaction (see Fig. 5). The possibility of a minor and competitive ionic mechanism was also considered.

Grohn and Vasiliu-Oprea (*18, 31, 32, 41*) studied the grafting of several vinyl monomers on polyamide, mainly polycaprolactam. [Their work was summarized in the Simionescu and Vasiliu-Oprea book (*13*).] In each case, the polymer was first purified by extraction, dissolution and reprecipitation. The grinding was performed at 0.42 fraction packing and by using 16 mm diameter steel balls, at an amplitude of 1.75 mm and at 1420 rpm. With vinyl chloride the air was eliminated from the reaction jar and the monomer was introduced at up to two

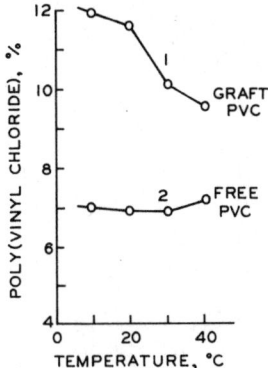

Fig. 6. Polymerization of vinyl chloride by vibromilling polyacrylonitrile. Effect of temperature on 1) graft PVC amount; 2) free homopolymer amount (13)

atm pressure (31, 32). The influence of the following parameters on the grinding synthesis was studied:

1. Powder size: two fractions with a particle size of 0.05–0.09 mm and 0.40–0.63 mm were used. The vibromilling was performed for 48 h at 10° C; The PVC content of the polymer was 6.57% for the first size particle and only 3.82% for the second, as the constant surface between polymer-monomer is less.
2. Moisture content: This parameter is important as moisture can induce a mechanochemically activated hydrolysis during vibromilling. The results, however, show that the moisture has only a minor effect on the graft content, indicating that the grafting process predominently occurs by homolytic bond cleavage. For the case of polycaprolactam, the PVC content after 36 h of vibromilling at 25° C was 35.55% for the reaction in dry medium and 36.08% in a wet medium. For the case of polyhexamethylenadipamide the corresponding values were respectively 38.39% and 38.37%.
3. Temperature: Mechanochemical degradation of polyamide by vibromilling exhibited a negative temperature coefficient. The grafting yield should thus increase on decreasing temperature. In fact, the reaction is temperature insensitive. From 10 to 40° C, only a modest decrease of grafting yield was observed. Homopolymerization is almost unaffected by temperature (see Fig. 6).

The reaction kinetics were studied on the total and separated components (graft and homopolymers) after 24–96 h of grafting at 10° C, see Fig. 7. The total yield increases almost linearly with time, and grafting is the dominant reaction. Extensive milling times has a negative effect, as the graft polymer is subjected to degradation. This was demonstrated by viscosity decreases with time.

The grafting of acrylonitrile and styrene on polycaprolactam was carried out in the liquid state (18, 31). The more important parameters were monomer concentration and the milling time. In order to separate the effect of total liquid and of monomer concentration, the reactions were also conducted in solution.

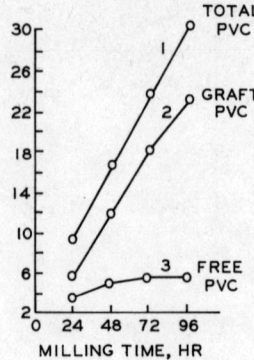

Fig. 7. Polymerization of vinyl chloride by vibromilling polycaprolactam. Composition at various millings 1) total PVC in the interpolymer, 2) graft PVC; 3) free PVC (13)

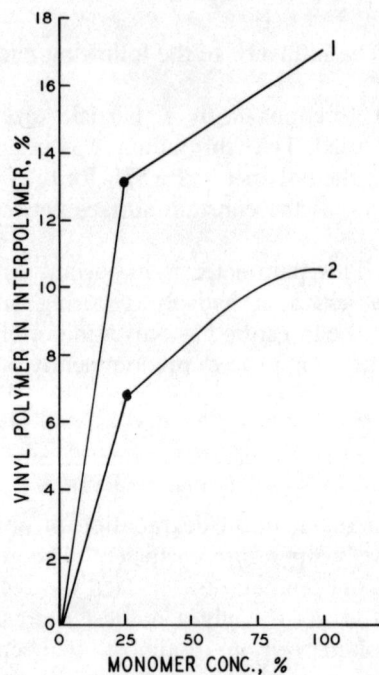

Fig. 8. Polymerization of vinyl monomers by vibromilling polyamide. Effect of monomer concentration on yielding 1) styrene; 2) acrylonitrile (13)

Monomer concentration in the range 25–100% had only a minor effect on the reaction, see Fig. 8, caused by milling 48 h at 20° C.

The effect of solvent chemistry was investigated on the system polyamide-acrylonitrile. Under the same conditions the acrylonitrile weight percent of the final product was 10.0, 12.7, and 13.7 respectively in the presence of methanol,

Table 6. Polymerization of vinyl monomers by vibromilling polycaprolactam. Composition change with milling time (13)

Monomer	Milling time h	Interpolymer composition %		Solvent insoluble fraction composition %		Homopolymer %
		Polyamide	Graft monomer	Polyamide	Graft monomer	
Vinyl chloride	24	90.64	9.36	94.33	5.67	3.61
	48	83.10	16.90	88.04	11.96	5.01
	72	76.02	23.98	81.62	18.36	5.60
	96	69.21	30.79	76.56	23.44	5.74
Acrylonitrile	24	100.00	0.00	100.00	0.00	—
	48	82.30	17.69	97.23		—
	72	72.59	27.41	87.65	12.35	—
	96	59.90	40.10	73.80	26.20	—
Styrene	24	90.58	9.42	92.49	7.51	1.91
	48	82.58	17.02	88.48	11.52	5.50
	72	75.69	24.31	83.06	16.94	7.37
	96	60.05	39.95	69.30	30.70	9.25

dioxane and acetene, respectively. The slightly less extent of reaction obtained using methanol as a solvent was explained by its action on the amide link (methanolysis).

For the case of acrylonitrile, there was an induction time of 24 h. This was attributed to the formation of "cyanide" radicals which are able to react with polyamidic macroradicals. The interpolymer is composed of two fractions one soluble in dimethyl formamide, whose properties are similar to polyacrylonitrile; the other, insoluble in this solvent, whose properties are similar to those of the polyamide. No homopolymer was observed. The presence of acrylonitrile on the graft polymer was demonstrated by IR.

For the case of styrene, the graft reaction was accompanied by homopolymerization. The presence of styrene on the graft polymer was also demonstrated by IR analysis.

The homopolymerization reaction proceeds through an intramolecular transfer reaction between macroradicals and monomers:

$$PA^{\cdot} + S \longrightarrow PA + S^{\cdot}$$

$$S^{\cdot} + nS \longrightarrow [S]^{\cdot}_{n+1}$$

Table 6 summarizes the results obtained with the three different monomers in mechanical reaction with polyamides (13).

Polyesters. Simionescu and coworkers continued their syntheses of heterochain polymers by investigation extensively the hehavior of polyethylene terephthalate by the vibromilling in the presence of different monomers. They achieved

Table 7. Polymerization of vinyl monomers by vibromilling poly(ethylene terephthalate). Composition change with milling time (33)

Monomer	Monomer state	Milling time h	Chemically grafted groups (—Cl, —CH, —OCOCH$_3$ respectively) %	Chemically linked	Homopolymer %
Vinyl chloride	Gas	12	7.40	13.04	1.94
		18	9.45	16.66	—
		24	13.30	23.45	2.99
Acrylonitrile	Liquid	12	1.61	3.18	2.92
		24	3.79	11.33	2.87
		36	6.20	22.30	1.20
		48	7.72	28.70	0.60
Vinyl acetate	Liquid	12	2.73	2.11	0.82
		18	5.10	3.89	1.46
		24	10.50	8.35	6.77

radically initiate graft and block polymerizations (33, 34) and, for the first time mechanically induced polycondensations, using diamines, and the synthesis of polychelates.

The activity of mechanically formed fragments in carrying out graft and block copolymerizations was studied both in gaseous and in liquid medium using vinyl chloride, acrylonitrile and vinyl acetate (33). The mechanochemical destruction of polyethylene terephthalate began after a long induction (12 h minimum). The grafting was carried out over intervals of 12–48 h at 18° C. The content of functional groups was determined by chemical and IR analysis, both before and after extraction with solvents to remove homopolymers. These data obtained are listed in Table 7. Higher degrees of grafting were obtained with gaseous monomers. In the liquid state, even acrylonitrile gave, for the same milling time (e.h. 24 h), much smaller amounts of chemically linked polymers. In all cases grafting processes predominate; homopolymerizations are only minor secondary reactions.

A peculiarity of copolymerization with gaseous vinyl chloride is the observation of high degrees of grafting even on short exposures to macroradicals formed by milling. This is probably due to chain transfer reactions during the initial stage of grafting.

The same technique was applied to a mixture of polyethylene terephthalate and acrylic acid (34). The polymerizations were followed by looking at the acid number of the product; the parameters studied were; time, temperature, and monomer content; see Fig. 9a, b, c. The hydrophilicity, the solubility of the copolymer in benzyl alcohol, aniline, and a mixture of phenol and $CHCl_3$ were increased by graft copolymerization.

Baramboim and coworkers (42) performed a low temperature copolymerization of a frozen polyethylene terephthalate suspension in acrylic acid, obtaining block copolymers and ternary acid resistant products.

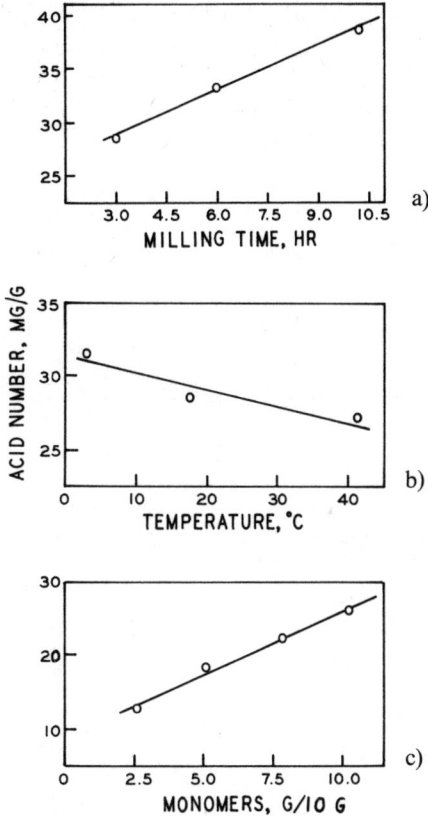

Fig. 9a–c. Polymerization of acrylic acid by vibromilling polyethylene terephthalate a) effect of milling time at 18° C on acid number of interpolymers; b) effect of temperature after 3 h milling on interpolymer acid number; c) monomer effect on interpolymer acid number after 3 h milling at 18° C (34)

As already described, the mechanochemical degradation of polyethylene terephthalate takes place mainly at the heteroatomic links (the weakest), while the breaking of —C—C— links is less pronounced. In wet media there is also a mechanochemically activated hydrolysis.

The graft and block reactions can be the following:

radical generation

...—O—(CH$_2$)$_2$OOC—⟨⟩—COO↓(CH$_2$)$_2$—OOC—⟨⟩—COO—(CH$_2$)$_2$—O—...

...—O—(CH$_2$)$_2$—OOC—⟨⟩—COO• + •CH$_2$—CH$_2$—OOC—⟨⟩—COO—(CH$_2$)$_2$—O—...

Grafting Initiation:

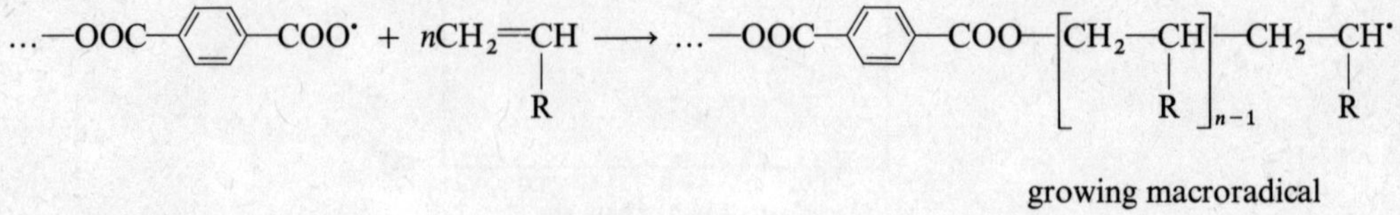

$$\ldots\text{—OOC—}\langle\text{C}_6\text{H}_4\rangle\text{—COO}^{\cdot} + n\text{CH}_2\text{=CH(R)} \longrightarrow \ldots\text{—OOC—}\langle\text{C}_6\text{H}_4\rangle\text{—COO—}[\text{CH}_2\text{—CH(R)}]_{n-1}\text{—CH}_2\text{—CH}^{\cdot}(\text{R})$$

growing macroradical

chain transfer:

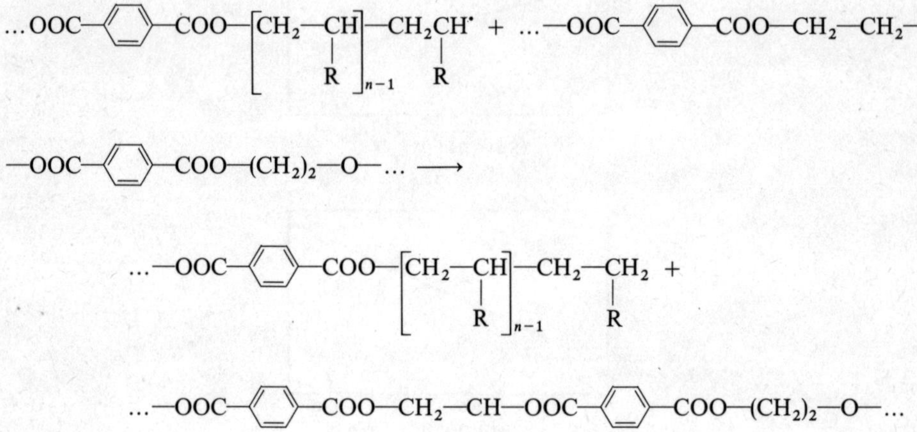

$$\ldots\text{OOC—}\langle\text{C}_6\text{H}_4\rangle\text{—COO—}[\text{CH}_2\text{—CH(R)}]_{n-1}\text{—CH}_2\text{CH}^{\cdot}(\text{R}) + \ldots\text{—OOC—}\langle\text{C}_6\text{H}_4\rangle\text{—COO—CH}_2\text{—CH}_2\text{—}$$

$$\text{—OOC—}\langle\text{C}_6\text{H}_4\rangle\text{—COO—(CH}_2)_2\text{—O—}\ldots \longrightarrow$$

$$\ldots\text{—OOC—}\langle\text{C}_6\text{H}_4\rangle\text{—COO—}[\text{CH}_2\text{—CH(R)}]_{n-1}\text{—CH}_2\text{—CH}_2(\text{R}) +$$

$$\ldots\text{—OOC—}\langle\text{C}_6\text{H}_4\rangle\text{—COO—CH}_2\text{—CH—OOC—}\langle\text{C}_6\text{H}_4\rangle\text{—COO—(CH}_2)_2\text{—O—}\ldots$$

grafting

$$\ldots\text{—OOC—}\langle\text{C}_6\text{H}_4\rangle\text{—COO—CH}_2\text{—CH—OOC—}\langle\text{C}_6\text{H}_4\rangle\text{—COO—(CH}_2)_2\text{—O—}\ldots$$

+ monomer or growing macroradical

$$\ldots\text{—OOC—}\langle\text{C}_6\text{H}_4\rangle\text{—COO—CH}_2\text{—CH(—CH}_2\text{—CH(R)—}[\text{CH}_2\text{—CH(R)}]_P\text{—CH}_2\text{—CH}^{\cdot}(\text{R}))\text{—OOC—}\langle\text{C}_6\text{H}_4\rangle\text{—COO—(CH}_2)_2\text{—O—}\ldots$$

growing macroradical

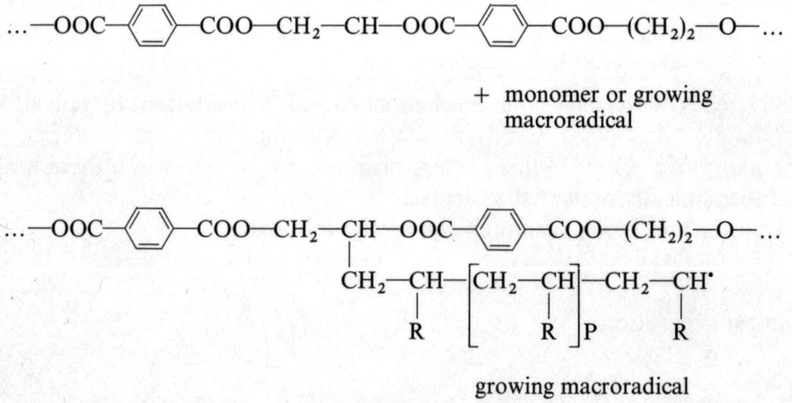

Table 8 summarizes the polymer-monomer systems investigated in mechanical synthesis in the solid state.

Table 8. Polymerization of monomers by vibromilling plastomers

Polymer	Monomer	Refs.	Temp. °C	Time	State
Poly(methyl methacrylate)	Vinyl chloride	(18–21)	0–55	36 h	Gas
	Acrylonitrile	(19)	40	36 h	Liquid
	Ethylene	(22)	−196 +20		Solid
	Acrylic acid		−196 +20		Solid
	Methylacrylate	(22)	−196 +20		Solid
	Acrylonitrile	(22)	−196 +20		Solid
	Styrene	(22)	−196 +20		Solid
	Butylene dimethacrylate	(22)	−196 +20		Solid
	Acrylic acid	(23)			Frozen
Poly(vinyl alcohol)	Acrylic acid	(24)	−78		Frozen
Poly(vinyl chloride)	Acrolein	(25)	23	4 h	Liquid[a]
Polyethylene	Methacrylamide	(26)	−30 −65	6 min	Solid
	Maleic anhydride	(26)	−30 −65	5 min	Solid
	Acenaphthylene	(26)	−30 −65	5 min	Solid
	Acrylic acid	(26)	−30 −65	5 min	Solid
	Acrylonitrile	(26)	−30 −65	3 min	Solid
Polybutylene dimethacrylate	Ethylene	(22)	−196 +20		Solvent
	Acrylic acid	(22)	−196 +20		Solvent
	Methylacrylate	(22)	−196 +20		Solvent
	Acrylonitrile	(22)	−196 +20		Solvent
	Styrene	(22)	−196 +20		Solvent
	Butylene dimethacrylate	(22)	−196 +20		Solvent
Polystyrene	Ethylene	(22)	−196 +20		Solvent
	Acrylic acid	(22)	−196 +20		Solvent
	Methyl acrylate	(22)	−196 +20		Solvent
	Acrylonitrile	(22)	−196 +20		Solvent
	Styrene	(22)	−196 +20		Solvent
	Butylene dimethacrylate	(22)	−196 +20		Solvent

Table 8 (continued)

Polymer	Monomer	Refs.	Temp. °C	Time	State
Polyacrylonitrile	Butadiene	(22)		30 h	Gas
	Vinyl chloride			180 h	Gas
Heterochain					
Polyamide	Acrylic acid	(27, 30)	− 78 −196	3 min	Solid
	Acrylonitrile	(18, 31, 41)	20	96 h	Liquid
	Styrene	(18)		96 h	Liquid
	Vinyl chloride	(31, 32)	10–40	96 h	Gas
Poly(ethylene terephthalate)	Acrylonitrile	(33)	18	48 h	Liquid
	Vinyl acetate	(33)	18	24 h	Liquid
	Vinyl chloride	(33)	18	24 h	Gas
	Acrylic acid	(34)	− 76 − 40	10 h	Liquid
Polyoxymethylene	Acrylonitrile	(28)	23	144 h	Liquid
	Methyl methacrylate	(2)	23	96 h	Liquid

[a] Standard mill.

Polycondensation. Polyethylene terephthalate carries active functional end groups (hydroxyl and carboxylic) capable of reacting with agents such as diamines. Simionescu, Vasiliu-Oprea and Neguleanu (33, 35, 37) thus initiated an investigation of a new type of synthesis: mechanochemical polycondensation. Their results are important as they widen the scope of mechanical synthesis, limited previously to radical initiation. The polycondensation reactions were carried out in the absence of air and moisture in a vibrating mill. The reaction was followed by measurement of polymer nitrogen content. Many factors influence the reaction: nitrogen content increased with duration of mechanical stress and with increasing amounts of diamines (Fig. 10a). The reaction is also influenced by the degree of mill packing (Fig. 10b). Particular attention must be put on the composition of the mill and balls. If porcelain was used, the results were corrected by the amounts of ash found on pyrolysis. It was found that ash content could be as high as 90%! As mechanical synthesis on freshly formed surfaces has been previously demonstrated, Simionescu and coworkers hypothesized the formation of a complex between the polycondensation products and the inorganic materials of the mill. To prove this hypothesis, IR spectra were recorded for the ball materials and the ash. The latter spectrum shows the characteristic bands for —C—NH and C—C bonds. The nitrogen content of the graft polymer increases with increasing reaction temperature (Fig. 10c). This is in contrast to a prime characteristic of mechanical synthesis. At lower temperatures, the macromolecular chain more efficiently stores mechanical energy as the viscosity is increased. The authors explained this anomaly by suggesting that the mechanical polycondensasation process take place at the end groups and that the activation is connected with chain mobility. At lower temperatures, (−5 to +40° C), the reaction is

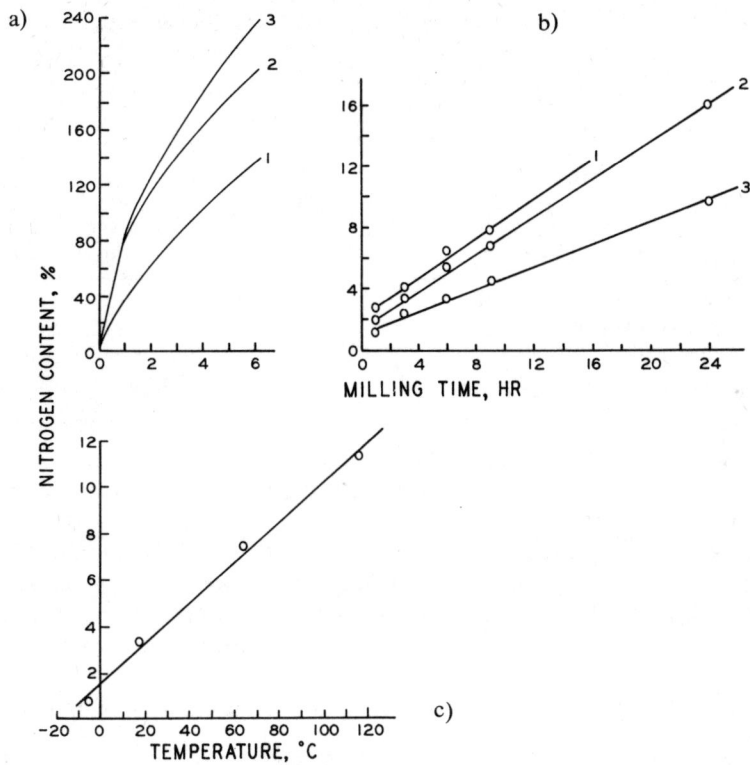

Fig. 10a–c. Mechanochemical polycondensation of polyethylene terephthalate and ethylendiamine: a) effect of milling time on nitrogen content of interpolymer at different polymer/condensating agent ratio. *1* 0.88; *2* 0.35; *3* 0.22 (*35*). b) effect of milling time on nitrogen content of interpolymer at different packing degree. *1* 0.2%; *2* 0.8%; *3* 3.2% (*37*). c) effect of temperature on nitrogen content of interpolymer (*37*)

probably only mechanical, whereas at higher temperatures, the polycondensation is initiated by thermal plus mechanical activation.

Other diamines besides ethylendiamine were employed in the course of this extensive investigation. Since most of the diamines selected are solids at room temperature they were reacted in solution or in the melt state (working at 75° C). The results can be summarized as follows:
1. Aromatic diamines are the most reactive; the reactivity of aliphatic diamines decreases with increasing molecular weight.
2. Diamines are more reactive than monoamines, especially if the amine group is in an aromatic para position.

The formation of copolymer by polycondensation reaction was proved by IR analysis. The spectra show absorption band at 1470, 1640, 2990, and 3100 cm^{-1} characteristic of —C—O—NH— groups, at 3300 cm^{-1} for imine groups and at 920 cm^{-1} for primary amine groups. From the point of view of chemical composition, the polycondensation products are composed by a fraction insoluble in

Table 9. Mechanochemical polycondensation of poly(ethylene terephthalate) with Polymer-monomer ratio, 0.11

Milling time h	Interpolymer % nitrogen	Interpolymer % PETP	ED	AA
3	10.90	72.19	19.64	8.17
6	14.50	70.58	25.91	15.92
12	16.72	61.18	29.85	19.07
15	21.62	41.58	36.80	14.87

PETP = Poly(ethylene terephthalate).
ED = Ethylene diamine.
AA = Adipic acid.

water (and in other common solvents) and a second soluble fraction. The second fraction shows an increase in electrical semiconduction and improved thermal stability.

Polycondensation reactions were also carried out using a mixture of ethylenediamine and adipic acid (35). IR techniques again were used to confirm the polymer composition. The results are summarized on Table 9. The chemistry of poly(ethylene terephthalate) mechanical polycondensation with diamines proceeds as follows:
1. Reaction of hydroxyl groups with amine units with formation of imino groups:

a) ...—COO—$(CH_2)_2$—OOC—⟨⟩—COO—$(CH_2)_2$—

—OOC—⟨⟩—COO—$(CH_2)_2$—OH + NH_2—$(CH_2)_n$—NH_2

↓ mechanical activation

...—COO—$(CH_2)_2$—OOC—⟨⟩—COO—$(CH_2)_2$—OOC—⟨⟩—COO—$(CH_2)_2$—

—NH—$(CH_2)_n$—NH_2 + H_2O

b) ...—COO—$(CH_2)_2$—OOC—⟨⟩—COO—$(CH_2)_2$—

—OOC—⟨⟩—COO—$(CH_2)_2$—OH + NH_2—$(CH_2)_n$—NH_2

$\xrightarrow{\text{mechanical activation}}$

...—COO—$(CH_2)_2$—OOC—⟨⟩—COO—$(CH_2)_2$—

—OOC—⟨⟩—COO—$(CH_2)_2$—O—$(CH_2)_n$—NH_2 + NH_3

2. Reaction of carboxylic groups on PETP support or formed by mechanical degradation and disproportionation with amine groups with formation of peptide groups. For example:

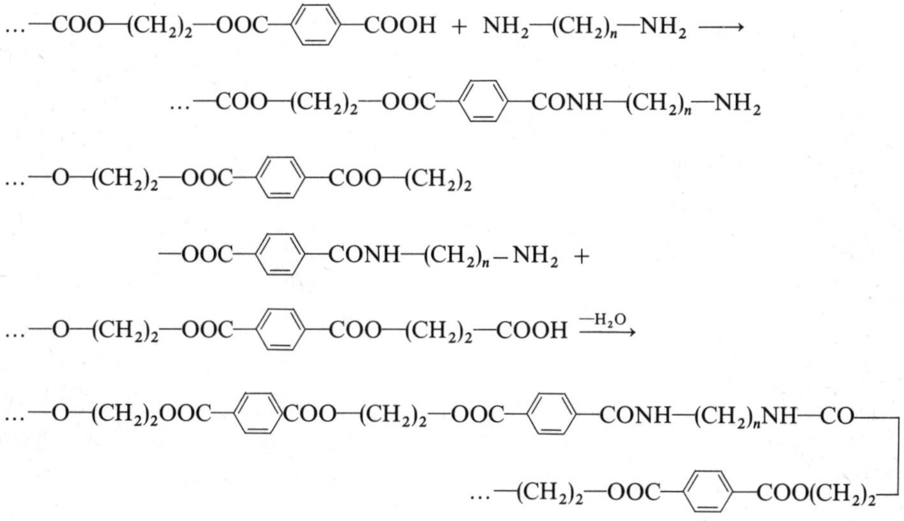

$$...-COO-(CH_2)_2-OOC-\langle\rangle-COOH + NH_2-(CH_2)_n-NH_2 \longrightarrow$$

$$...-COO-(CH_2)_2-OOC-\langle\rangle-CONH-(CH_2)_n-NH_2$$

$$...-O-(CH_2)_2-OOC-\langle\rangle-COO-(CH_2)_2$$

$$-OOC-\langle\rangle-CONH-(CH_2)_n-NH_2 +$$

$$...-O-(CH_2)_2-OOC-\langle\rangle-COO-(CH_2)_2-COOH \xrightarrow{-H_2O}$$

$$...-O-(CH_2)_2OOC-\langle\rangle COO-(CH_2)_2-OOC-\langle\rangle-CONH-(CH_2)_nNH-CO-$$

$$...-(CH_2)_2-OOC-\langle\rangle-COO(CH_2)_2-$$

3. If adipic acid is also added to the blend the reactions are more complicated, as the monomers can react with each other. The final products are:

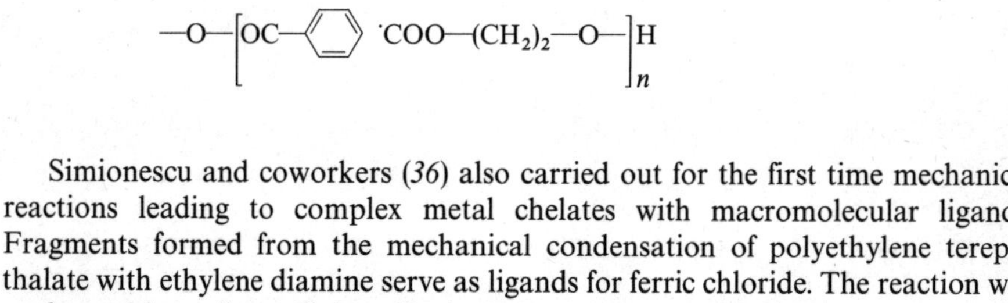

$$H_2N-(CH_2)_n-NHCO-(CH_2)_4-COO(CH_2)_2$$

$$-O-\left[OC-\langle\rangle-COO-(CH_2)_2-O-\right]_{n-1} OC-\langle\rangle-COO(CH_2)_2OH$$

$$H_2N-(CH_2)_2-NHCO-(CH_2)_4-CONH-(CH_2)_2-NH-(CH_2)_2$$

$$-O-\left[OC-\langle\rangle\ COO-(CH_2)_2-O-\right]_n H$$

Simionescu and coworkers (36) also carried out for the first time mechanical reactions leading to complex metal chelates with macromolecular ligands. Fragments formed from the mechanical condensation of polyethylene terephthalate with ethylene diamine serve as ligands for ferric chloride. The reaction was performed in a vibromill (amplitude 4 mm, 1475 rpm at 18° C).

The influence of the milling time and of diamine and ferric chloride content were studied, see Fig. 11.

The polychelate is thermostable, possesses semiconductor properties and has a reduced solubility.

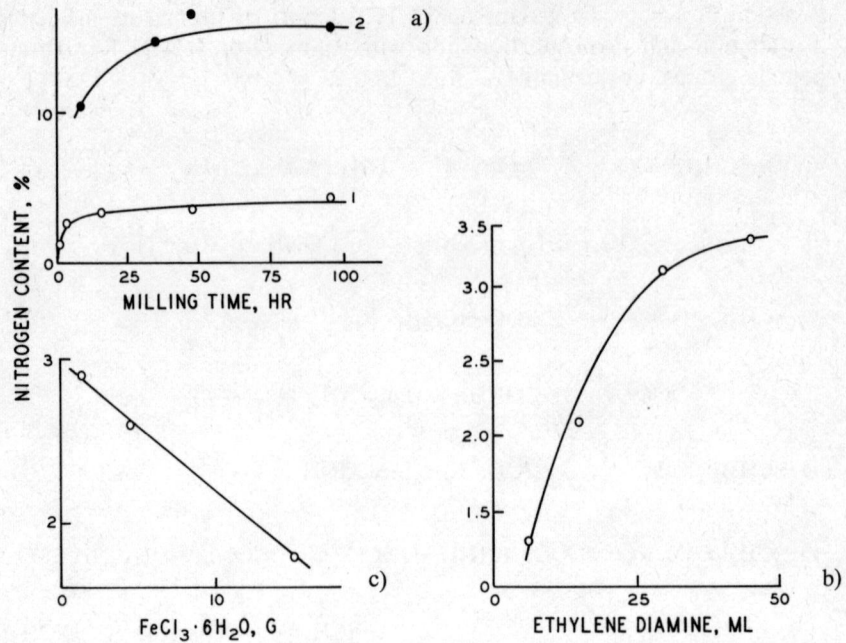

Fig. 11a–c. Mechanochemical synthesis of polychelates: a) effect of milling time on nitrogen content of interpolymer *1* 10 g PETP, 30 g ETDA, 6.5 g FeCl$_3$. 6 H$_2$O; *2* γ polycondensation without ferric salt (*36*). b) effect of ETDA content on nitrogen content of interpolymer (10 g PETP, 6.5 g FeCl$_3$. 6 H$_2$O, milling time 15 h) (*36*). c) Effect of FeCl$_3$. 6 H$_2$O content on nitrogen content of interpolymer (10 g PETP, 30 g ETDA, milling time 3 h) (*36*)

The chemistry of polychelate formation proceeds as follows:
1. Polycondensation of polyester and diamine

...—CH$_2$—CH$_2$—OOC—⟨⟩—COO—(CH$_2$)$_2$—OOC—⟨⟩—COO—CH$_2$—... $\xrightarrow{\text{mechanical degradation}}$

...—CH$_2$—CH$_2$—OOC—⟨⟩—COO· + ·CH$_2$—CH$_2$—OOC—⟨⟩—COO—CH$_2$— ⟶ disprop.

...—CH$_2$—CH$_2$—OOC—⟨⟩—COOH + CH$_2$=CH—OOC—⟨⟩—COO—CH$_2$

+ NH$_2$—CH$_2$—CH$_2$—NH$_2$ $\xrightarrow{\text{polycondensation}}$

—CH$_2$—CH$_2$—OOC—⟨⟩—CONH—CH$_2$—CH$_2$—NH—CO—⟨⟩—COO—CH$_2$—CH$_2$—

2. Complex formation

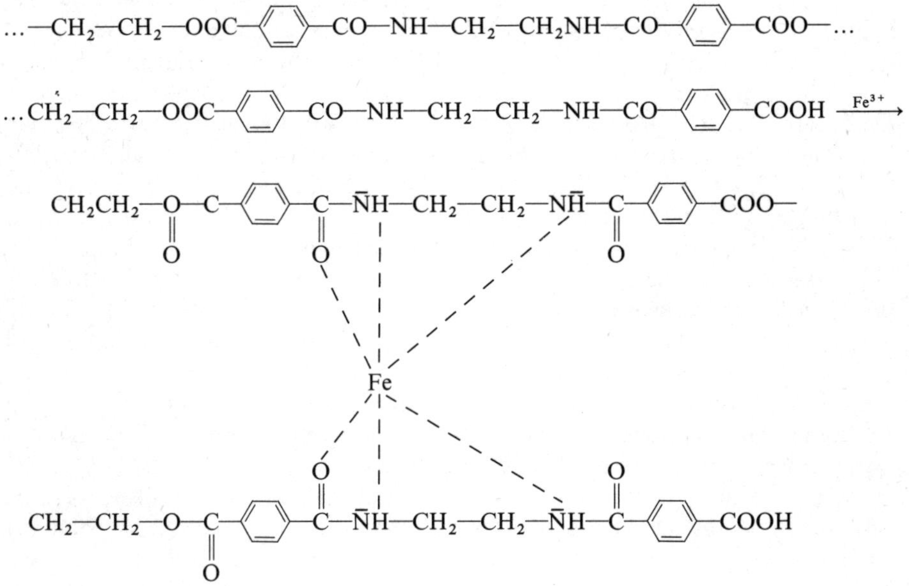

Besides this and a tautomeric form, the chelate could also have the form:

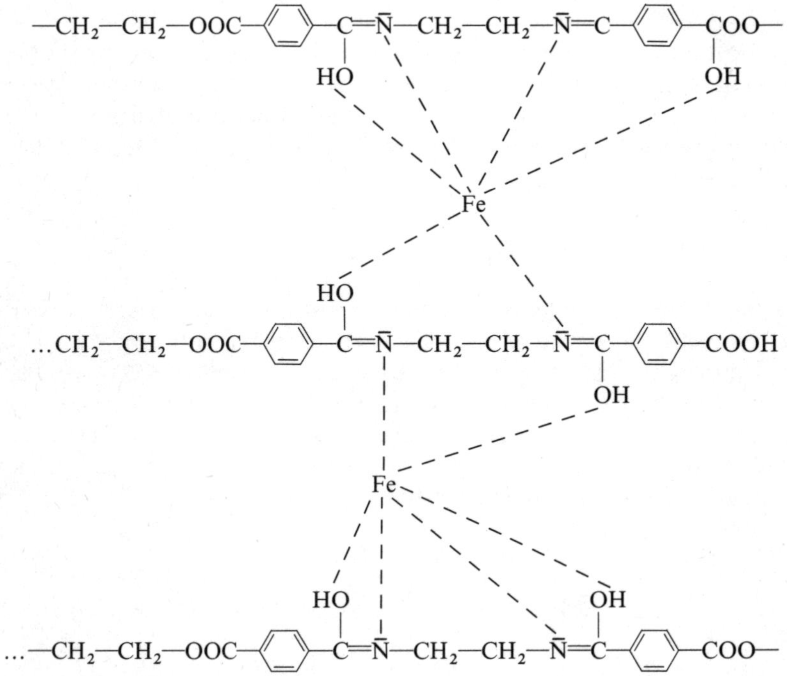

2. Rubbery State

Mastication in the rubbery state has been one of the most studied methods among the possibilities of making block and graft copolymers by mechanochemistry. A large body of literature covers this subject; particularly detailed studies were performed by the English researchers of NRPRA, Natural Rubber Producers Research Association (*1, 12*). An extensive patent literature has also been authored by these same researchers, see for example, Brit. pat. 828, 895; 832, 193; 832, 700; 836, 053; 842, 205; 851, 731. These studies followed the observation that natural rubber is greatly changed (broken down) by cold mastication. This early mechanochemical reaction by mastication is now known to be generally applicable to polymers, natural and synthetic, which exhibit viscoelasticity below their decomposition temperatures.

Syntheses have been carried out on polymer-polymer, polymer-monomer, and polymer-filler systems. The properties of the products obtained can vary widely according to chemical structure and the conditions of mastication (temperature, mixing intensity, presence and nature of radical acceptors and stabilizers, atmosphere, solvents and ratio of blend components).

With polymer blends, it is possible to influence the composition of final product by changing the relative viscosities during the reaction, by changing molecular weights, by use of solvents, or by selecting a particular temperature. Note that (1) mechanical reactions have a negative coefficient; (2) degradation is reduced at high temperature in simple flow; and (3) viscosity depends only slightly on the temperature below T_g.

In the case of polymer-monomer systems, it is possible to regulate the reaction by adding monomer at different stages or in different order with reactions being sufficiently rapid for commercial processes.

The mastication equipment most commonly employed is standard rubber instrumentation such as roll mills, internal mixers, extruders or laboratory devices modeled on them (*e.g.*, a single-rotor internal masticator described by Wilson and Watson (*43*), the model improved by Kargin and coworkers (*11*) and the Brabender plastograph).

a) Polymer-polymer Systems

Elastomers. Many experiments have been performed with natural rubber, using either synthetic elastomers or plastomers as the second polymer.

As stated previously, the intrinsic viscosity of natural rubber decreases after mastication under nitrogen, while the Huggins' interaction constant, k', increases, suggesting branching, without leading, however, to gel formation. Branching can be explained by a hydrogen transfer reaction between rubber molecules and polymeric radicals, resulting from the mechanical scission (*44*)

$$\sim \underset{\underset{H}{|}}{\overset{\overset{CH_3}{|}}{C}}=C-CH_2^{\cdot} + \sim \underset{\underset{H}{|}}{\overset{\overset{CH_3}{|}}{C}}=C-CH_2-CH_2-\overset{\overset{CH_3}{|}}{C}\sim \; ----- \; \sim \underset{\underset{H}{|}}{\overset{\overset{CH_3}{|}}{C}}=C-CH_3 + \sim \underset{\underset{H\;H}{|\;|}}{\overset{\overset{CH_3}{|}}{C}}=C-\overset{\overset{CH_3}{|}}{C^{\cdot}}-CH_2-\overset{\overset{CH_3}{|}}{C}\sim$$

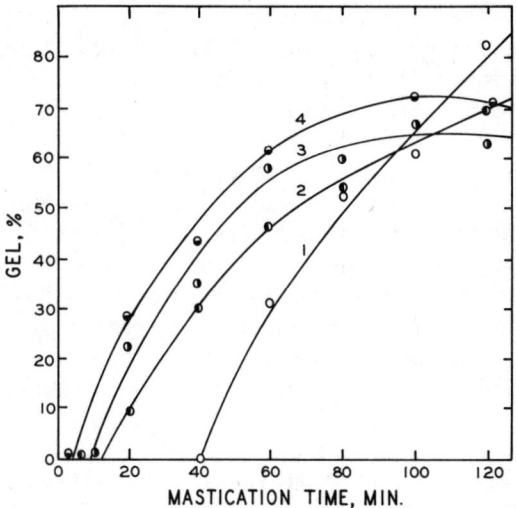

Fig. 12. Mastication of natural rubber-neoprene systems. Effect of mastication time on gel content. Neoprene content *1* 100%; *2* 75%; *3* 50%; *4* 25% (*44*)

and by combination of the secondary radicals with primary type. In contrast, the most common synthetic elastomers gel during mastication in inert atmospheres. This tendency to gelification can be explained by the presence of functional groups, which can activate the double bond or of pendent vinyl groups which are more susceptible than the internal double bonds to reaction with polymeric radicals to give gel.

Mastication under nitrogen for a blend of natural and synthetic rubbers can result in the formation of block and graft copolymers as well as gel. A typical example is the mastication of natural rubber with neoprene (*44*). The electron withdrawing chlorine substituent enhances the reactivity of the double bond to radical attack, causing gelification during mastication. The degradation was carried out in laboratory masticator (*44*) under nitrogen at 30° C and 76 rpm. The gel production on milling is illustrated in Fig. 12. Neoprene remained wholly soluble for up to 40 min of mastication after which gel was rapidly produced until it comprised 80% of the elastomer after two hours. Increasing amount of natural rubber, up to 75%, reduced the time to gel point to about five minutes. Gel increase after the gel point was slower than with neoprene alone. The gel also forms more rapidly at higher shear rates. Evidence of copolymer formation was obtained by solubility characteristics and by selective precipitation. Similar results were obtained during mastication of natural rubber with polybutadiene (*45*), nitrile rubber (*45*), and styrene-containing rubber (*45–47*). Comastication of natural rubber and styrene rubber was carried out in a screw extruder at 35–40° C (*48*). Decreasing of temperature and an inert atmosphere increased polymer modification. In the mastication of blend of natural rubber and styrene rubber in an inert atmosphere the chains of the former rupture preferentially (*46*).

The macroradicals of natural rubber react with those of the styrene elastomer, due to the presence of the very reactive 1,2 pendent vinyl groups in the latter. This mechanism leads to a structure where the styrene rubber forms a gel network with grafted branches of natural rubber.

Similar results were obtained by shear blending of two synthetic elastomers (*45*). The formation of a block or graft copolymer during the process of mixing butadiene rubber (SKB) and styrene-containing rubber (SKS-30A) was postulated by Slonimskii and Reztsova (*49*). They claimed that the anomalies observed in the dependences on composition of the mechanical properties of a mixture of two mutually insoluble rubbers after vulcanization may be reduced by increasing the part played by the mechanical mixing (inert atmosphere, reduction of radical acceptors, intensity of mechanical action).

Mastication of natural rubber with elastomers which do not give gel allows the synthesis of soluble graft or block copolymers, depending on the tendency for chain rupture and transfer reactions.

Graft polymers were obtained by mastication of a 60/40 blend of natural rubber and chlorosulphonated polyethylene (*1*). From 10–55% natural rubber was obtained as side chains. Grafting presumably proceeds by transfer of chlorine atoms to the rubber radicals to give grafting sites for combination with rubber radicals. Soluble linear polymers were also obtained by mastication for 50–180 min under nitrogen for a blend 50/50 of natural rubber and a polyurethane rubber (Vulcaprene A) (*1*).

Block copolymers based on nitrile rubber and on epoxy and phenolic resins and on polystyrene (*50–54*) have been intensively studied in Russia. The generated block copolymers were investigated by turbidimetric and IR methods. Thermomechanical experiments were also run on fractions. As may be seen from Fig. 13, fractions which combine the properties of the polymers (Curves 2, 3, and 4) were obtained together with fractions characteristic of the raw rubber (Curve 1) and of the resin (Curve 5). The copolymer is soluble in solvents which are typical for both components. Solubility studies on the products showed that for any given ratio of the original components, 15 to 20% of the resin combines with the rubber. The properties of the block copolymer, however, depend on the initial ratio of components: nitrile rubber confers elasticity and the phenolic resin processability.

Beniska and Staudner (*55*) described a different method for the grafting of vinyl polymers on plasticized rubber. In a first stage they performed a degradation at 20–25° C in the presence of oxygen and hydroperoxide groups. The hydroperoxides are decomposed by heat (at 90, 100, and 120° C) with formation of macroradicals which initiate polymerization of monomers. It is claimed that with increasing time of plasticizing and temperature, the conversion of monomers increased.

Grafting of polytetrafluoroethylene (Teflon) on polydimethylsiloxane has been carried out by milling with the aim of improving solvent (benzene) resistance, tensile strength, and tear resistance (*5*). As organosilocon rubber does not readily undergo scission by mechanical action, it was necessary to add benzoylperoxides to the blend before mixing. Apparently the process is related to the scission of polydimethylsiloxane initiated by peroxide. The macroradicals formed may enter into reaction with the radical products of mechanical scission of polytetrafluoro-

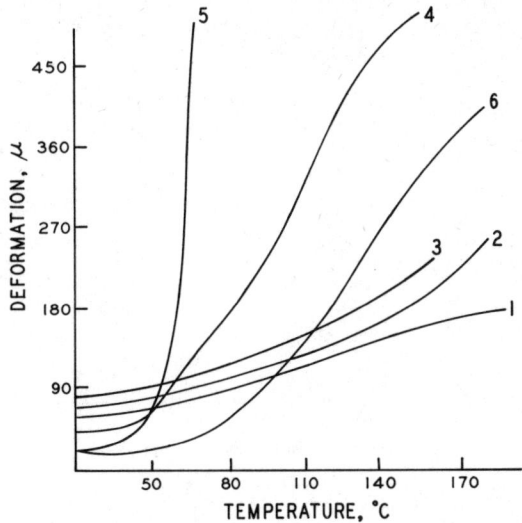

Fig. 13. Dependence of deformation on temperature for a block polymer 6, based on novolac resin 17 and nitrile ubber, ratio 1/1, and for fractions *1–6* (*53*)

Table 10. Mechanical synthesis of interpolymers by mastication of polymer blends

			Refs.
Natural rubber	Plus	neoprene	(*44–45*)
Natural rubber		polybutadiene	(*45*)
Natural rubber		nitrile rubber	(*45*)
Natural rubber		styrene rubber	(*45–48*)
Natural rubber		chlorosulphonated polyethylene	(*1*)
Natural rubber		polyurethane rubber	(*1*)
Polybutadiene		styrene rubber	(*49*)
Neoprene		styrene rubber	(*45*)
Nitrile rubber		epoxy resin	(*50–51*)
Nitrile rubber		phenolic resin	(*53–50*)
Nitrile rubber		polystyrene	(*53*)
Polytetrafluoroethylene		polydimethylsiloxane	(*5*)
Butyl rubber		polyethylene	(*45*)

ethylene with formation of branched and block copolymers. The systems studied are summarized in Table 10.

Plastomers. The production of resins (poly[vinyl chloride], polystyrene, and poly[styrene-co-acrylonitrile]) with relatively high toughness has been one of the most important aims of industry. This can be achieved by modifying a rigid chain with small amounts of elastomers. The best results have been obtained by the use of block and graft copolymers.

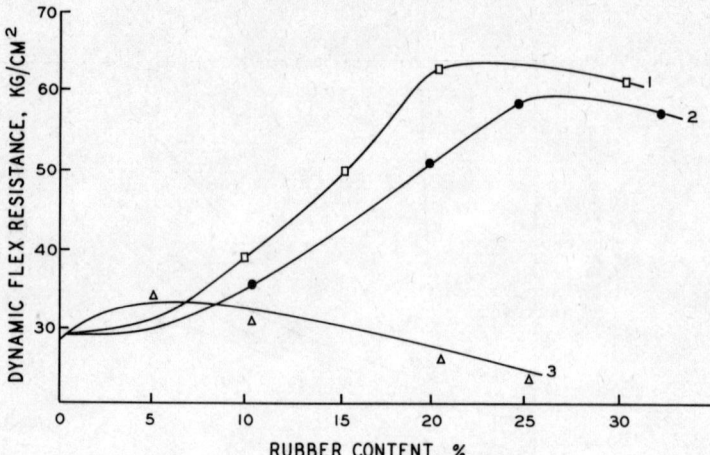

Fig. 14. Mastication of polystyrene — synthetic rubbers systems. Effect of the amount and type of rubber on dynamic flex resistance. *1* styrene rubber (SKS 30); *2* nitrile rubber (SKN 18); *3* nitrile rubber (SKN 40) (5)

Berlin and coworkers (5, 56) desired to obtain a material with an increased mechanical strength. They carried out a plasticization of bulk and emulsion polystyrene molecular weight 80000 and 200000 respectively at 150–160° C, with polyisobutylene, butyl rubber, polychloroprene, polybutadiene, styrene rubber (SKS-30) and nitrile rubber (SKN 18 and SKN 40). The best results were obtained with the blends polystyrene-styrene rubber and polystyrene-nitrile rubber. An increase of rubber content above 20–25 % was not useful, as the strength properties were lowered. An increase in the content of the polar comonomer, acrylonitrile, prevents the reaction with polystyrene and decreases the probability of macroradical combination. This feature lowers the strength, see Fig. 14. It was also observed that certain dyes acts as macroradical acceptors, due to the mobile atoms of hydrogen of halogens in the dye, AX:

$$...-CH-CH_2^{\cdot} + AX \longrightarrow ...-CH-CH_2 x + A^{\cdot}$$

Thus in coloring polymers by mastication it should be considered that dyes of different classes, even in small amounts, may affect the material properties, see Table 11.

The possibility of increasing poly(vinyl chloride) toughness has been investigated by many workers (52, 57–59). The effect of modification depends markedly on the compatibility of the initial substances, their polarity, and structure.

Poly(vinyl chloride) was also masticated with neoprene in an inert atmosphere at 145° C for 30 min (60), giving an easily flowing powder of low bulk density. Goto and coworkers (61) blended poly(vinyl chloride), poly(methyl methacrylate) and polystyrene in an open roll after investigating the degradation condition for each component to find the optimum combination of the component.

Table 11. Mastication of polystyrene-rubber system. Effect of dyes on interpolymer strength (5)

Additive	Amount, g/100 g resin	Flex resistance kg cm/cm^2	
		Dynamic	Static
No additive	2.0	53.3	930
Violet vat dye	0.2	47.6	870
Pigment "yellow"	0.5	27.5	—
"Bordeaux" lac dye	0.2	30.9	850
Crystalline iodine	0.1–0.2	29.9	943
Channel black	2.0	57.6	835

b) Polymer-monomer Systems

Elastomers. The British researchers at NRPRA were the first to study intensively and systematically the possibility of performing mechanical syntheses by masticating a blend of rubber and monomer, utilizing the radicals from chain scission to initiate the polymerization of monomers (1). Le Bras and coworkers (62–64) however, had described many years earlier the possibility of changing the properties of rubber by mixing with monomers. They claimed to have synthesized a new substance called "anhydride rubber" by treating rubber with maleic anhydride in a mixing machine (62). They underline, also, that acrylic acid and acrylonitrile bring about the same type of transformation of rubber. This effect was attributed to short links formed by the monomers between or across the rubber molecules. The same authors continued their study after the fundamental work of Watson and coworkers. Their results supported the British researchers conclusion concerning the radical mechanism of synthesis (65–67). Le Bras investigated also the role played by oxygen in the process. His conclusion (66) was that the reaction between rubber and maleic anhydride is accelerated by oxygen concentration reduction.

Mechanical synthesis by cold mastication of rubber and monomers depends on the reaction condition (monomer concentration, temperature, solvent concentration, atmosphere, presence of transfer agents, or catalyst) and on the physical and chemical properties of the rubbers, the monomers and the product interpolymers. A critical factor is the shear stress developed in the system rather than instrumentally-defined shear rates. The degree of reaction of polymer and consequently also the concentration of free macroradicals depends on stress. As a consequence, the influence of the above parameters may be connected to their influence on the viscosity of the reaction medium since an increase in viscosity causes an increase in stress at constant shear rate.

Watson and coworkers (68, 69) studied in details the mechanical reaction of systems of natural rubber plus poly(methyl methacrylate), using a laboratory masticator at 76 rpm, in a nitrogen atmosphere at 15° C. Prior to mastication, the deproteinized crepe rubber was extracted with acetone, imbibed with monomer and allowed to homogenize for 16 h, in the absence of light. It must be underlined that, with limited mastication and reaction heat dissipation, the actual rubber temperature was much higher; the maximum recorded was about 50° C and

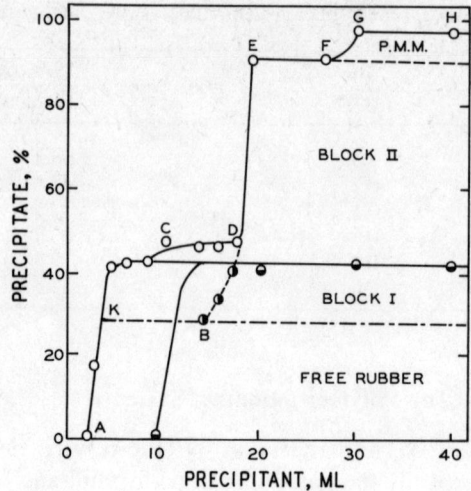

Fig. 15. Separation of the polymeric products obtained after masticating natural rubber containing 38.5% methyl methacrylate to 97% conversion: (○) fractional precipitation by methanol from 10 ml of 1% benzene solutions; (◐) fractional precipitation by acetone; (●) fractional precipitation by methanol after removal of free rubber; (—·—) fraction extracted by petroleum ether

depended on shear intensity and the material properties. It was thus difficult to investigate independently the effect of shear rate as an increase of rotor speed increases the thermal softening of the rubber by frictional heat. Fractional precipitation and extraction methods allowed separation of the product in two homopolymers and two different interpolymer fractions, see Fig. 15 (68).

Influence of Condition for Mechanical Reactions. For natural rubber, Fig. 16a shows the effect of monomer concentration on polymerization. Although small amount of monomer had little effect on bulk viscosity, the initiation and the polymerization rate were affected; higher concentrations did produce disproportionately marked plasticization and consequently long induction periods for mastication, even though the polymerization rate was nearly the same (69). This behavior is due to the autocatalytic nature of the polymerization and directly related to the fast increase in bulk viscosity during the consumption of monomer and the formation of the more rigid plastomer chain. As the reaction proceeds, the bulk viscosity of the system increases rapidly and, at constant shear rate, the concentration of polymeric free radicals formed by mechanical scission increases correspondingly. The effect of temperature is illustrated in Fig. 16b. The reaction possesses the usual negative coefficient, typical of mechanochemical reactions (69). Dilution by solvent produced an increased induction time and a less marked reduction in the polymerization rate, Fig. 16c (69). The presence of transfer agent (tert-dodecylmercaptan) reduced the rate of conversion (Fig. 16d) (68). The presence of catalyst only reduced the induction time, without affecting the polymerization rate (69).

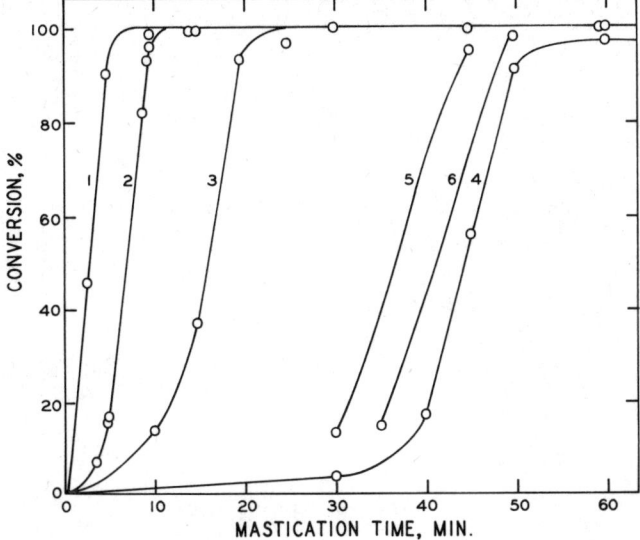

Fig. 16a

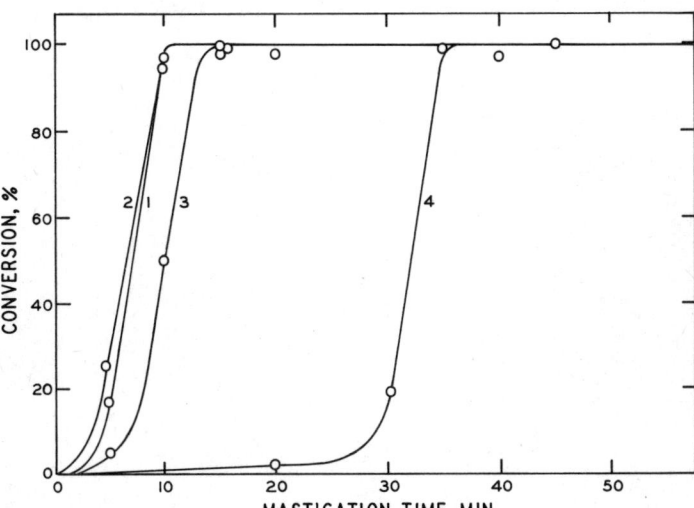

Fig. 16b

Fig. 16a–d. Polymerization of methyl methacrylate by natural rubber mastication: a) effect of time, monomer concentration and catalyst on monomer conversion. Initial monomer concentration *1* 23.8%; *2* 38.5%; *3* 48.5%; *4* 55.6%; *5* 55.6% + 1% benzoyl peroxide; *6* 55.6% + 1% bisazoisobutyronitrile (*69*). b) effect of temperature on monomer conversion at 76 RPM (initial monomer concentration 38.5%) *1* 15° C; *2* 15° C at 360 RPM; *3* 25° C; *4* 35° C (*69*). c) effect of solvent on monomer conversion. *1* 23.8% methyl methacrylate. *2* 38.5% methyl methacrylate. Vol. 2 ml of concentrations: *3* 2:1 methyl methacrylate: C_6H_6. *4* (◐) 1:1 methyl methacrylate: C_6H_6; (◉) 1:1 methyl methacrylate: CCl_4. *5* (●) 1:2 methyl methacrylate: C_6H_6. d) effect of transfer agent on monomer conversion (initial monomer concentration 38.5%). 0, 0.2, 0.5, 2.0, and 5.0 ml tert.-dodecylmercaptan per 100 ml monomer respectively. The original reference gives the viscosity at each point for the polymeric products measured on the Wallace Rapid Plastimeter (*68*)

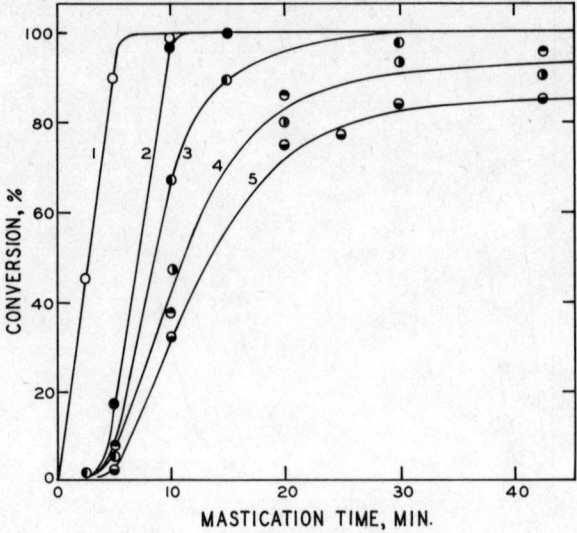

Fig. 16c

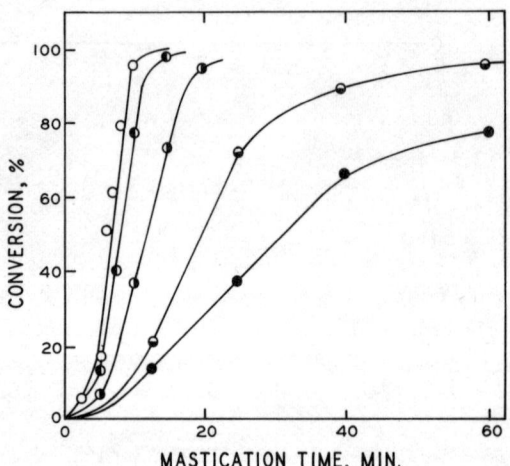

Fig. 16d

Influence of Rubber Properties. The physical and chemical properties of rubbers, monomers and interpolymers must be considered together, as in the main stage of the reaction all of these are present and the properties of the mixture at any moment determine the subsequent reaction.

The chemical nature of the rubber determines which bonds are the weakest and are therefore more likely to be ruptured during mastication by the statistical concentration of mechanical energy about such bonds. An increase in the degree of asymmetry, an increase in the stiffness and the packing density of macromolecules facilitate mechanical scission; resonance stability will influence the

Table 12. Polymerization of styrene by poly(methyl acrylate) mastication (30% of monomer). Effect of rubber molecular weight on interpolymer composition (11)

Initial intrinsic viscosity, dl/g	Free poly(methyl acrylate)	Free polystyrene	Block copolymer
5.85	4	7	89
1.81	19	5	77
1.01	31	8	61
0.51	53	5	42

reactivity of polymeric free radicals once formed, opposing the tendency to disproportionate and favoring recombination reaction; the presence of active groups will favor the formation of graft copolymer and so on. The physical properties of the initial rubbers influence principally the early stages of reaction. High bulk viscosity and molecular weight are positive factors in using a rubber as initiator. The molecular weight influences the amount of free elastomers content at the end of reaction. This was observed for both natural rubber-methylmethacrylate (68) and for poly(methyl acrylate)-styrene (Table 12) (11).

Mastication was also carried out on crosslinked rubber (natural, styrene, nitrile, butyl and neoprene) swollen with different monomers. The reaction was run in four stages. It was possible to combine up to 70% of monomers by allowing the crumb-like mastication products to take up additional monomer during the reaction. The conversion was found to be dependent on the vulcanization recipe (Table 13) (70). In no case was significant free rubber formed during mastication. Methyl methacrylate and methacrylic acid gave high conversions, no conversion of styrene higher than 16% was found. It was also not possible as yet to demonstrate the reaction initiation site (main chain, crosslink or rubber-filler bond scission).

Figure 17 (69) shows the mechanical behavior of different synthetic elastomers. They were found to be generally less efficient than natural rubber in promoting polymerization because of reduced stress during mastication due to greater softening by monomer addition. Nitrile rubber crumbed with methyl methacrylate, styrene and acrylonitrile.

Influence of Monomer Properties. Polymerization depends primarily on the chemical and physical properties of monomer. First of all the monomer must be sufficiently reactive for radical polymerization. Vinyl acetate failed to polymerize during mastication because of the low reactivity of alkyl radicals towards this monomer. Gelation occurs on mastication with monomers which give very active free radicals capable of reacting with the low activity groups of natural rubber. The effect of monomer composition was investigated by adding small amounts to avoid an influence on reaction products (68). A high compatibility of rubber and monomer leads to slower reaction, again due to viscosity reduction. This factor facilitates the movement of chains over each other under the action of shear and thus reduces rupture. The softening effect of different monomers is illustrated on Table 14 (11). Reaction conditions are the same as described for the

Table 13. Polymerization of monomers by vulcanized rubber mastication. Effect of vulcanization recipes (1, 70)

Vulcanization recipes, parts per hundred parts rubber by weight	Monomers	Conversion %			
		Stage 1	Stage 2	Stage 3	Stage 4
Pale crepe, 3 dicumyl peroxide; 50 min at 140°C	Methyl methacrylate Methacrylic acid	87 94	93 82	97 —	94 —
Acetone-extracted pale crepe, 5 zinc oxide, 1 stearic acid, 1 phenyl-β-naphthylamine, 0.7 Santocure, 2.5 sulfur; 30 min at 140°C	Methyl methacrylate	5	20	23	—
Pale crepe, 50 phenol-formaldehyde resin, 5 zinc oxide, 1 stearic acid, 1 phenyl-β-naphthylamine, 0.7 Santocure, 2.5 sulfur; 30 min at 140°C	Methyl methacrylate Styrene Methacrylic acid	68 10 94	85 30 80	95 — —	— — —
Pale crepe, 20 phenol-formaldehyde resin, 3 dicumyl peroxide 50 min at 140°C	Methyl methacrylate Styrene Methacrilic acid	94 16 79	95 33 91	94 — —	95 — —
Pale crepe, 50 phenol-formaldehyde resin, 5 hexamethylene tetramine, 5 zinc oxide, 1 stearic acid, 1 phenyl-β-naphthylamine, 0.7 Sontocure, 2.5 sulphur, 30 min at 140°C	Methyl methacrylate	43	57	24	—
Pale crepe, 20 cresol-formaldehyde resin, 2 hexamethylene tetramine, 3 dicumyl peroxide; 50 min at 140°C	Methyl methacrylate	91	95	12	—
Smoked sheet, 40 lamp black, 1 phenyl-β-naphthylamine, 3.5 dicumyl peroxide; 50 min at 140°C	Methyl methacrylate	52	—	96	96
Smoked sheet, 40 lamp black, 5 zinc oxide, 1 stearic acid, 1 phenyl-β-naphthylamine, 0.7 Santocure, 2.5 sulfur; 30 min at 140°C	Methyl methacrylate	22	—	16	—
Neoprene GN, 5 zinc oxide, 4 magnesium oxide, 0.5 stearic acid; 30 min at 140°C	Methyl methacrylate	36	—	25	—
Butyl 200, 25 furnace black, 5 zinc oxide, 1 stearic acid, 1 tetramethyl thiuram-disulphide, 0.5 mercaptobenzothiazole, 1 sulfur; 40 min at 153°C	Methyl methacrylate	55	—	39	31
Butyl 200, 25 furnace black, 5 zinc oxide, 1 stearic acid, 1 tetramethyl thiuram-disulphide, 0.5 mercaptobenzothiazole, 1 sulfur; 40 min at 153°C extracted	Methyl methacrylate	95	—	—	—

Table 13. Polymerization of monomers by vulcanized rubber mastication. Effect of vulcanization recipes (1, 70)

Vulcanization recipes, parts per hundred parts rubber by weight	Monomers	Conversion %			
		Stage 1	Stage 2	Stage 3	Stage 4
Hycar OR-15, 30 high abrasion furnace black, 10 zinc oxide, 0.5 tetramethyl thiuramdisulfide, 1.5 sulfuric; 30 min at 153°C	Methyl methacrylate	49	—	—	—
Polysar S 50, 5 zinc oxide, 2 stearic acid, 1 Nonox HFN, 1 Santocure, 1.75 sulfur; 40 min at 153°C	Methyl methacrylate	31	—	28	25
Polysar S 50, 5 di-t-butyl-peroxy-butane; 40 min at 153°C	Methyl methacrylate	33	—	63	30
Polysar S 50, 5 di-t-butyl-peroxy-butane; 40 min at 153°C extracted	Methyl methacrylate	96	—	—	—

MMA methyl methacrylate
MA methacrylic acid
S styrene

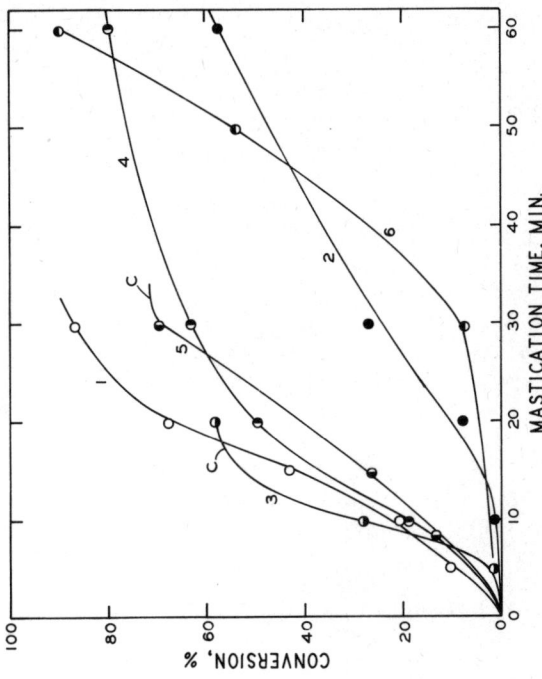

Fig. 17. Polymerization of monomers by synthetic rubber mastication. 1 13.8% methyl methacrylate in neoprene. 2 13.6% chloroprene in neoprene. 3 13.8% methyl methacrylate in polybutadiene-styrene. 4 13.6% chloroprene in polybutadiene-styrene. 5 14.5% styrene in polybutadiene-arylonitrile. "c" indicates that the rubber became a crumb at approximately the arrowed time of mastication. 6 11% methyl methacrylate in rubber — from Table X in original

Table 14. Plasticity of natural rubber containing 0.66 ml/g of different monomers in Wallace Plastimeter units (*11*)

Monomer	Plasticity in WP units
Methyl methacrylate	78
Ethyl methacrylate	66
n-Butyl methacrylate	61
Nonyl methacrylate	59
Lauryl methacrylate	75
Methacrylic acid	68
Divinyl benzene	60
Styrene	45
Methyl vinyl ketone	98
N-Vinyl pyrrolidone	98

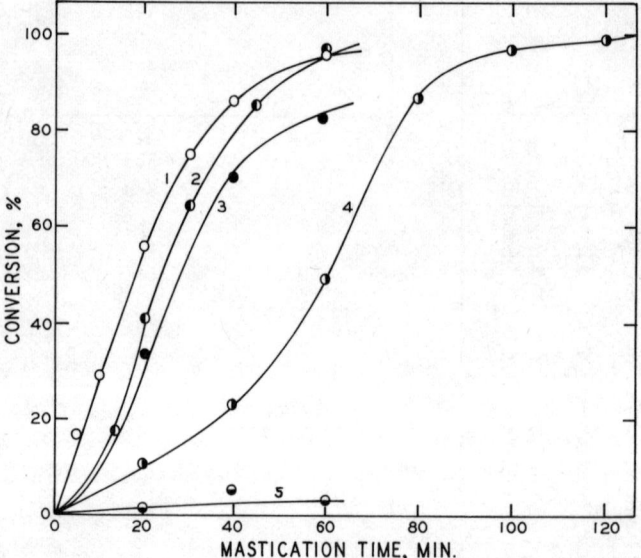

Fig. 18. Polymerization of styrene by natural rubber mastication. Effect of time, monomer concentration and temperature on monomer conversion. *1* 13.2% styrene, 15° C; *2* 23.3% styrene, 15° C; *3* 23.3% styrene, 25° C; *4* 37.7% styrene, 15° C; *5* 37.7% styrene, 25° C (*69*)

system natural rubber-methylmethacrylate. Figure 18 illustrates the effect of monomer concentration and temperature on styrene polymerization (*69*). The reaction rate is slower than in vibromilling. It could be useful to underline that in this last case the tests were run at temperatures well below the glass transition temperature of the resins and the system viscosities are nearly the same.

The interpolymers obtained with different elastomers vary widely; methyl methacrylate and styrene give a soluble product whereas chloroprene, acrylo-

Table 15. Polymerization of monomers by mastication of elastomers

		Refs.
Natural rubber	Maleic anhydride	(62–67)
Natural rubber	acrylic acid	(62, 64, 69)
Natural rubber	acrylonitrile	(62, 64, 69)
Natural rubber	methyl methacrylate	(68, 69)
Natural rubber	styrene	(69)
Natural rubber	chloroprene	(69)
Natural rubber	acrylonitrile	(69)
Natural rubber	methacrylic acid	(69)
Natural rubber	vinyl acetate	(69)
Natural rubber	methyl acrylate	(69)
Natural rubber	ethyl acrylate	(69)
Natural rubber	divinyl benzene	(69)
Neoprene	methyl methacrylate	(69)
Neoprene	chloroprene	(69)
Styrene rubber	methyl methacrylate	(69)
Styrene rubber	chloroprene	(69)
Styrene rubber	maleic anhydride	(71)
Nitrile rubber	styrene	(69)
Nitrile rubber	maleic anhydride	(72)
Butyl rubber	methyl methacrylate	(69)
Butyl rubber	styrene	(69)
Polyisobutylene	methyl methacrylate	(69)
Polyisobutylene	maleic anhydride	(71, 72)
Polybutadiene	maleic anhydride	(73)
Polyesteramide rubber	methyl methacrylate	(69)
Polyesteramide rubber	styrene	(69)

nitrile and methacrylic acid produce a gel containing the polymerized monomer. Extensive investigations were carried out on maleic anhydride (71–74) for the reinforcement of rubber and because of the peculiar reaction. Table 15 summarizes the most important results obtained using natural and synthetic rubbers with different monomers.

Influence of Interpolymer Properties. As stated earlier, the physical and chemical properties of interpolymers markedly influence the reaction rate after the induction period. If the monomer present yields a polymer comparable in viscosity with the initial mixture the rate of scission will not accelerate. For example, the polymerization rate of chloroprene on mastication with natural rubber does not increase as markedly with conversion (69), see Fig. 19, as with methyl methacrylate and styrene. The reason is the chloroprene-rubber system remained elastic and softer than the original rubber.

Experiments were also performed with the aim of polymerizing a mixture of two monomers (69). The reaction rate and the composition of the graft copolymer conform to the rules of mechanochemical synthesis and radical copolymerization. If the two monomers have almost equal reactivities, the composition of the copolymer is approximately that of the initial monomer mixture (Fig. 20). The

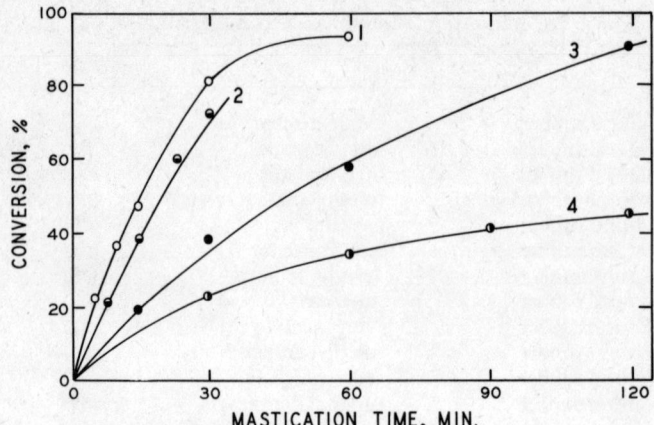

Fig. 19. Polymerization of chloroprene by natural rubber mastication. Effect of time, monomer concentration and temperature on monomer conversion. *1* 24.2% chloroprene, 15° C; *2* 24.2% chloroprene, 25° C; *3* 39.0% chloroprene, 15° C; *4* 49.0% chloroprene, 15° C (*69*)

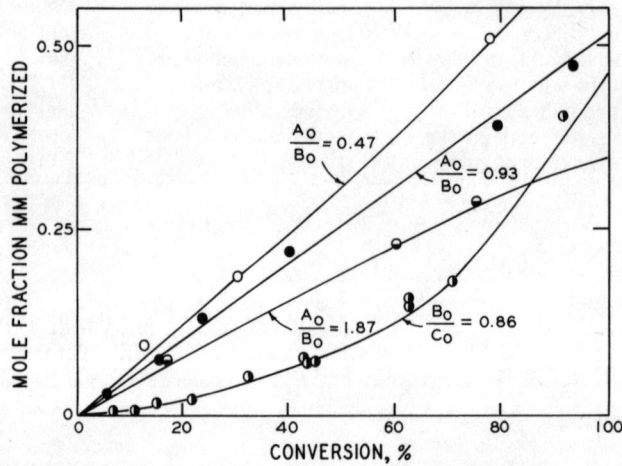

Fig. 20. Composition of copolymers formed by rubber mastication: (A_0, B_0, C_0) initial molar fractions of styrene, methyl methacrylate, and chloroprene, respectively. (Full lines) calculated compositions based on r_1 and r_2 values of 0.50 and 0.44 for styrene and methyl methacrylate and of 0.083 and 6.12 for methyl methacrylate and chloroprene (*69*)

reaction rate is also intermediate between those of the separate monomers. If the two mechanically-activated monomers have different reactivities, the two monomers largely polymerize separately. The reaction rate will depend on the polymer properties for the monomer which polymerizes first. As an example, in the system chloroprene-methyl methacrylate (*69*), the more active chloroprene polymerizes at first: the reaction rate is low as polychloroprene is a flexible chain

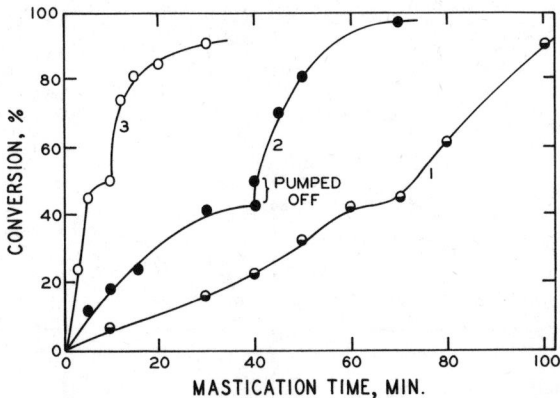

Fig. 21. Polymerization of methyl methacrylate and chloroprene by rubber mastication (69). *1* 23.8% methyl methacrylate and 24.2% chloroprene added initially. *2* 24.2% chloroprene polymerized, then 23.8% methyl methacrylate. *3* 23.8% methyl methacrylate polymerized, then 24.2% chloroprene

and the second monomer behaves as an inert solvent, reducing the viscosity. The reaction rate increases in the second part of the reaction when an appreciable amount of poly(methyl methacrylate) is produced, see Fig. 21. Two monomers added consecutively can give two different results. If the chloroprene is added first: the initial reaction rate is faster, as there is no plasticizing effect of the second monomer. If the methacrylate is added first, the reaction rate of chloroprene (second part of Curve 3, Fig. 21) is higher because of the effect of rigid poly(methyl methacrylate) on the initiation rate.

The presence of a comonomer has, in certain cases, a marked influence on polymerization rate. For example, the mastication of natural rubber in the presence of maleic anhydride, even with small concentrations of the latter, about 5%, leads to accelerated polymerization of styrene monomer (*11*) either because of its high reactivity in the propagation step of heterochain copolymerization and/or because of a hardening effect. This reaction is discussed later.

Reaction Mechanism. Angier and Watson (*68*) carried out a detailed study on the reaction mechanism of the system natural rubber-methyl methacrylate. The structures of the interpolymers were investigated, after fractional precipitation and extraction, by measurements of composition, osmotic molecular weight, and ozonolysis to degrade rubber segments and by isolation of poly(methyl methacrylate) fragments, the autoxidation of the rubber segments, and viscosity measurements on the uncombined rubber. By these analyses it was concluded that reaction produced two homopolymers and two block copolymers, see Fig. 15. The two block copolymers are formed by one plastomer segment between two rubber segments. The Block I, which is formed in the initial stages of reaction, is composed of long rubber and short plastomer segments. This is consistent with the low initiation rate, the preferential rupture of the longer rubber molecules, and growth of plastomer chains less influenced by the Trommsdorff gel effect.

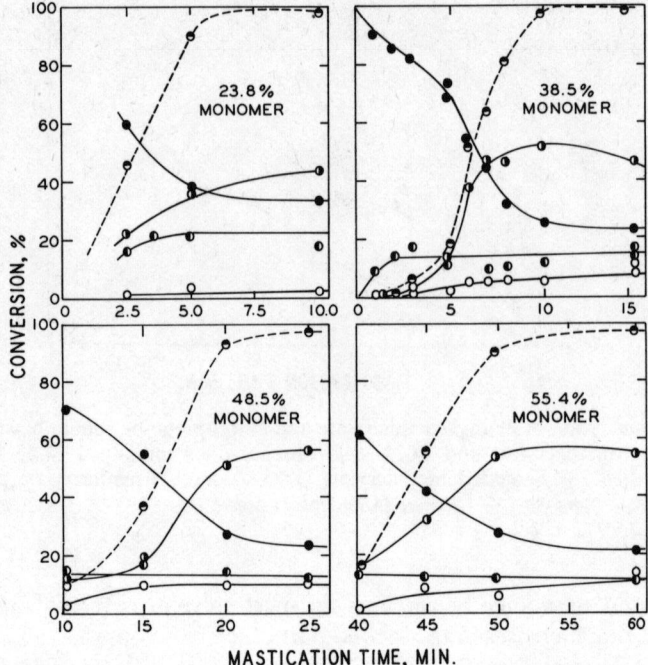

Fig. 22. Polymerization of methyl methacrylate by natural rubber mastication (68). (☉) extent of polymerization; (●) free rubber; (◐) Block I (the numbers refer to the percentage of PMM in the samples); (◑) Block II; (○) free PMM. The points correspond to a range of PMM concentrations as specified in the original reference

The Block II, which is formed during the period of rapid polymerization, is a larger fraction than Block I, and possesses short rubber and long plastomer segments. This is consistent with the decreasing termination rate caused by the gel effect; even so the initiation rate is higher. It must be, also, emphasized that at the end of the reaction there is appreciable free rubber with a molecular weight higher than the limiting molecular weight reached on cold mastication of rubber. This is explained by the incomplete degradation of the original rubber during mastication in blend with monomer and by the effect of shear on the block copolymers with preferential rupture of the block chains near the link between rubber and plastomer. This last point was demonstrated by milling the isolated block copolymers; the former feature was studied by controlling the capability of the separated free rubber to be ruptured and to initiate the polymerization of monomer. The amount of free rubber can be reduced by performing the reaction by successive addition of monomers. The ratio of the two block copolymers and the two homopolymers depends on reaction condition: by increasing the initial monomer concentration the Block II fraction increases, see Fig. 22, whereas a temperature increase slightly increases the free rubber content. These general effects were also obtained on solvent addition although the homo-poly(methyl methacrylate) was reduced almost to zero and the content of the block copolymers decreased. The addition of solvent changed also the structure of the block copolymers (an increase of plastomer content).

A detailed study was performed for the system rubber-maleic anhydride. The following reaction products were proposed on the basis of the known reactivities:

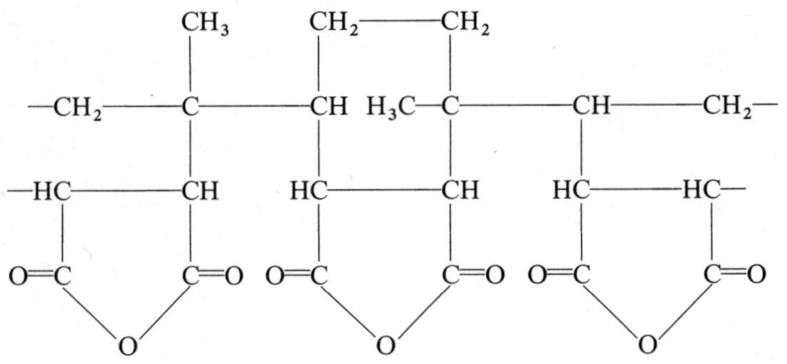

(1)

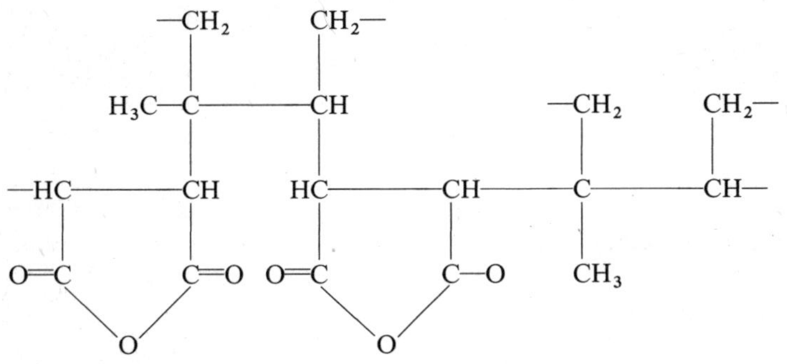

(2)

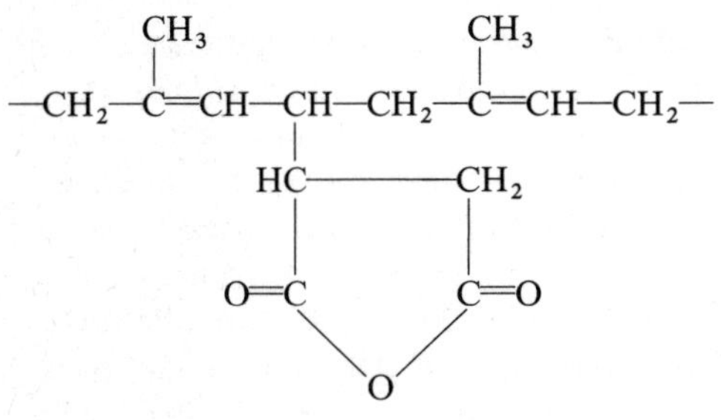

(3)

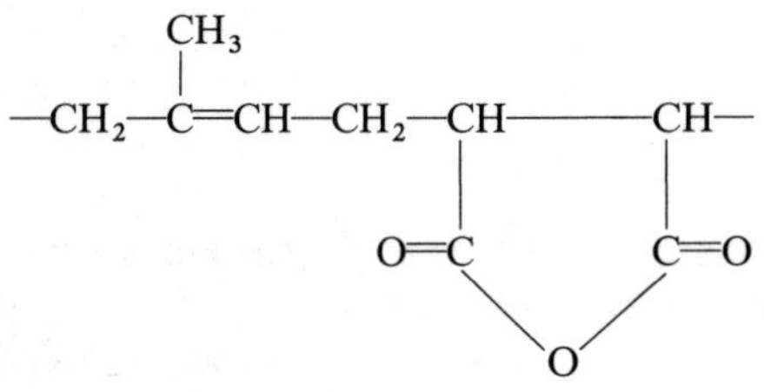

(4)

Table 16. Modification of rubbers properties by masticating with vinyl monomers (1, 75)

Elastomer	Monomer		Tensile strength kb/cm²	Elongation at creak %	Modulus at extension kb/cm²	
					100%	300%
Polybutadiene (SKB)	—		149	428	27	99
	Styrene		154	564	21	71
			178	380	49	149
	Methyl methacrylate	(9%)	163	420	31	112
		(17%)	177	484	31	104
		(23%)	172	324	49	
Poly(butadiene-co-styrene) (SKS 30A)	Methyl methacrylate	(9%)	290	820	15	39
	—		254	764	12	54
Natural rubber	—		142	610	5	16
	Methyl methacrylate	(11%)	166	465	10	46
		(21%)	154	365	17	99
		(36%)	171	285	62	—
	Styrene	(12%)	121	450	10	38
		(21%)	142	430	20	62
	Methacrylic acid	(9%)	261	245	18	16
		(20%)	124	200	50	41

It was concluded that most probably maleic anhydride combines with the primary macroradicals (Scheme 4). Secondary radicals may be found by reaction with unsaturated groups of the rubber to add more monomers units along the chain and to produce a network.

Properties of Block and Graft Copolymers. Properties of a virgin rubber can be changed greatly by mechanical reaction. Acrylic acid, at 25% concentration, gave a hard rubber and at 40% an unyielding material that stopped the masticator (69). The acrylates and acrylonitrile produced a rubber crumb similar in appearance to the product with maleic anhydride. Samples containing more than 40% acrylamide gave a fine powder insoluble in benzene (69). The mechanochemical binding of maleic anhydride makes it possible to carboxylate the polymers, as anhydride rings are easily opened by moisture. In general maleic anhydride leads to an increase in the stiffness, strength, adhesion to metals, hydrophilic nature and possibility of crosslinking with polyvalent metal oxides (73). Table 16 shows modification of natural and synthetic rubbers by masticating specifically with vinyl monomers (1, 75).

c) Plastomers and Natural Polymers

As already stated, mastication reactions are not limited to elastomers but can be extended to all polymers in the viscoelastic state. It is thus interesting to note that before the fundamental study of Watson and coworkers on cold rubber mastication, Reid (76) had already found that when vinyl polymers and monomers are subjected to mechanical deformation both degradation and polymerization occur simultaneously. The concept is that the state and not the polymer nature is the prime factor determining the formation of macroradicals. Indeed the NRPRA researchers carried out mechanical syntheses on a range of nonrubber polymers (77–82). Their results are summarized in Table 17 and Fig. 23 (77). Generally, the reactions conform to the rules for rubbers but with faster reaction rates, due to higher bulk viscosities, and give more rigid polymeric polymerization products. Polypropylene was grafted by Russian researchers. Romanov et al. (83) masticated atactic polypropylene at 0° C and at 300 rpm while cold styrene vapor carried in a nitrogen stream was passed at 1 cc/h per gram of polypropylene. Baramboim (84) and coworkers masticated the same polymers with 5–15% of maleic anhydride at 20° C for 15–45 min and obtained a graft product. He also (85) studied the system polyamide-poly(methyl methacrylate). The systems poly(methyl methacrylate)-styrene and polystyrene-methyl methacrylate were studied by Watson and coworkers (77). The composition of these interpolymers is shown in Fig. 24 (77). In the polystyrene-methyl methacrylate system, 53% of the original polymer remained as homopolymer at the end of the reaction. Homopoly(methyl methacrylate) was formed during the early stages of reaction (40 and 91% of the polymerized monomer after 80 and 95% of conversion respectively). The amount of block polymer reached a maximum of only 42% of the product.

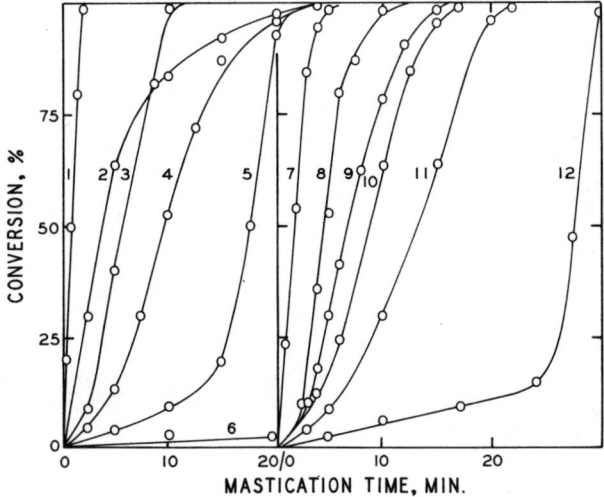

Fig. 23. Polymerization of monomers in masticating polystyrene and polymethyl methacrylate. Curves *1–6*: 1 ml methacrylic acid, styrene, methyl methacrylate, ethyl acrylate, acrylonitrile, and vinyl acetate, respectively, in 3 g polystyrene. Curves *7–12*: 2 ml methacrylic acid, methyl methacrylate, acrylonitrile, ethyl acrylate, styrene, and vinyl acetate, respectively, in 3 g polymethyl methacrylate. The limiting viscosity numbers for points along Curves *2* and *3* are given in the original

Table 17. Polymerization of monomers by plastomer mastication (82)

Polymer	Monomer	%	Mastication time min	Monomer convers. %
Poly(vinyl acetate)	Methyl methacrylate	25.3	10	98
	Vinyl acetate	12.0	20	96
	Styrene	36.3	15	96
	Acrylonitrile	20.1	35	94
	Di-isobutene	20.0	40	89
	Methacrylic acid	26.1	2	96
	Acrylic acid	18.9	10	98
Poly(vinyl formal)	Methyl methacrylate	21.0	25	97
	methacrylic acid	39.5	5	93
	Styrene	25.2	35	89
	Vinyl acetate	25.7	45	86
	Acrylonitrile	22.2	20	98
	Di-isobutene	23.2	55	89
	Octyl acrylate	19.8	30	94
	Hexyl methacrylate	23.5	25	91
	Methoxyethyl methacrylate	26.4	20	92
	Methacrylonitrile	23.5	10	98
	Vinyl caproate	27.9	25	95
	Vinyl behanoate	25.0	30	88
Poly(vinyl butyral)	Styrene	25.0	15	97
	Methyl methacrylate	24.3	10	97
	Vinyl acetate	23.1	20	85
	methacrylic acid	29.0	1	97
	Acrylonitrile	21.1	20	95
	Di-isobutene	24.7	70	86
Poly(n-butyl-methacrylate)	Methyl methacrylate	25.0	20	98
	Styrene	24.3	30	97
	Vinyl acetate	21.0	55	89
	Methacrylic acid	28.7	15	99
	Di-isobutene	19.0	40	87
	Allyl acrylate	23.5	15	94
	Octyl acrylate	25.6	15	92
	n-Butyl acrylate	24.7	10	98
	Ethyl acrylate	26.4	10	96
Poly(methyl methacrylate)	Methyl methacrylate	40.4	7.5	88
	Styrene	33.2	20	95
	Acrylonitrile	34.8	10	95
	Methacrylic acid	34.0	5	86
	Acrylic acid	32.1	10	99
	Ethyl acrylate	36.8	10	97
	Divinyl benzene	37.5	2.5	86
	Vinyl acetate	37.5	30	97
	Allyl methacrylate	40.2	15	96
	β-Ethoxyethyl methacrylate	31.4	5	93
	2-Chloroethyl methacrylate	40.1	10	97
	Ethyl methacrylate	37.0	5	97
	Ethylene dimethacrylate	40.9	10	96

Table 17 (continued)

Polymer	Monomer	%	Mastication time min	Monomer convers. %
	Allyl acrylate	38.3	5	92
	n-Butyl acrylate	37.7	5	97
	Methyl acrylate	34.8	5	98
	Methacrylonitrile	30.2	5	87
	Methyl isopropenyl ketone	35.5	15	97
	N-Vinyl pyrrolidone	41.0	10	95
	2-Vinyl pyrrolidone	39.6	20	86
	Vinylidene chloride	35.0	10	99
Polystyrene	Methyl methacrylate	26.3	10	99
	Styrene	23.4	15	98
	Methacrylic acid	34.2	5	98
	Acrylic acid	32.1	7.5	97
	Di-isobutylene	17.0	25	86
	Acrylonitrile	21.1	20	98
	Allyl acrylate	25.4	25	96
	Allyl methacrylate	28.7	20	96
	n-Butyl acrylate	25.0	20	95
	n-Butyl methacrylate	25.0	15	99
	Isobutyl methacrylate	25.8	20	
	2-Chloroethyl methacrylate	25.4	20	96
	β-Ethoxyethyl methacrylate	26.0	20	94
	Ethyl acrylate	25.7	20	99
	2-Ethylhexyl acrylate	26.4	20	96
	Ethyl methacrylate	25.0	15	98
	Lauryl methacrylate	23.5	25	98
	Nonyl methacrylate	22.5	25	89
	N-Vinyl pyrrolidone	26.4	20	97
	Octyl acrylate	25.3	20	94
	2-Vinyl pyridine	28.4	20	87
Poly(vinyl chloride)	Acrylonitrile	23.0	50	97
	Allyl acrylate	38.0	20	96
	Allyl methacrylate	40.0	15	96
	n-Butyl acrylate	37.0	20	95
	n-Butyl methacrylate	24.0	10	98
	Isobutyl methacrylate	27.0	20	96
	Divinyl benzene	37.0	20	99
	2-Choroethyl methacrylate	25.0	40	96
	Ethyl acrylate	26.0	20	94
	β-Ethoxyethyl methacrylate	25.0	45	96
	2-Ethylhexyl methacrylate	36.0	25	94
	Ethyl methacrylate	37.0	20	93
	Methyl methacrylate	40.0	20	99
	Methylisopropenylketone	26.0	35	96
	Methyl vinyl ketone	25.0	60	94
	N-Vinyl pyrrolidone	43.0	20	94
	Styrene	33.0	20	92
	Vinylpyridine	40.0	20	93
Poly(vinylidene chloride)	Acrylonitrile	27.0	45	95
	Allyl acrylate	25.0	20	96
	Allyl methacrylate	25.0	20	92

Table 17 (continued)

Polymer	Monomer	%	Mastication time min	Monomer convers. %
	n-Butyl acrylate	25.0	20	97
	n-Butyl methacrylate	25.0	25	96
	Isobutyl methacrylate	25.0	20	99
	2-Chloroethyl methacrylate	26.0	25	92
	Di-isobutene	20.0	35	89
	β-Ethoxyethyl methacrylate	29.0	20	95
	Ethyl acrylate	23.0	25	98
	2-Ethylhexyl acrylate	25.0	30	95
	Ethyl methacrylate	23.0	25	96
	Lauryl methacrylate	26.0	30	91
	Methyl methacrylate	27.0	20	98
	Methyl vinyl ketone	25.0	45	86
	Nonyl methacrylate	28.0	20	89
	N-Vinyl pyrrolidone	30.0	20	96
	Octyl acrylate	28.0	25	95
	Styrene	25.0	30	93
	Vinyl acetate	21.0	45	90
	2-Vinyl pyridine	28.0	45	90
Poly(n-vinyl pyrrolidone)	Acrylonitrile	25.0	55	96
	Allyl acrylate	22.0	25	96
	Allyl methacrylate	25.0	20	92
	Butyl methacrylate	26.0		89
	Isobutyl methacrylate	25.0	30	87
	n-Butyl acrylate	23.0	30	89
	2-Chloroethyl methacrylate	28.0	35	90
	β-Ethoxyethyl methacrylate	25.0	25	94
	Ethyl acrylate	25.0	35	96
	Ethyl methacrylate	24.0	30	90
	Lauryl methacrylate	26.0	30	85
	Methacrylic acid	30.0	15	99
	Methyl methacrylate	25.0	20	97
	Methacrylonitrile	23.0	15	92
	Nonyl methacrylate	22.0	35	87
	n-Vinyl pyrrolidone	28.0	35	99
	2-Vinyl pyrrolidone	24.0	30	92
	4-Vinyl pyridine	23.0	25	93
	2-Vinyl 5-ethyl pyridine	25.0	20	89
	2-Vinyl 5-methyl pyridine	22.0	30	90
Poly(styrene-co-butadiene)(85/15)	Acrylonitrile	22.0	20	95
	Allyl acrylate	25.0	20	90
	n-Butyl acrylate	25.0	25	91
	n-Butyl methacrylate	24.0	20	95
	Isobutyl methacrylate	23.0	25	90
	Di-isobutene	21.0	35	85
	β-Ethoxyethyl methacrylate	27.0	25	90
	Ethyl acrylate	25.0	20	98
	Ethyl methacrylate	28.0	25	96
	2-Ethylhexyl methacrylate	26.0	25	91
	Lauryl methacrylate	23.0	45	90

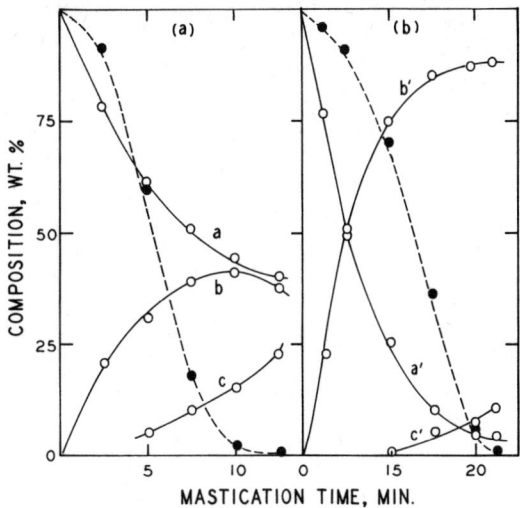

Fig. 24. Composition during the mastication of initially (a) 24% methyl methacrylate in polystyrene and (b) 38% styrene in polymethyl methacrylate. Curves *a*, *b*, and *c*: Free polystyrene, block polymer, and free polymethyl methacrylate, respectively. Curves *a'*, *b'*, and *c'*: Free polymethyl methacrylate, block polymer and free polystyrene, respectively. (— —) % unpolymerized monomer

In the poly(methyl methacrylate)-styrene system, less than 7% of the original polymer remained as homopolymer at total conversion (77). Over 85% of the product was non-branched, single-segment block copolymer. The difference for these two systems is in part due to the higher molecular weight of the initial poly(methyl methycrylate) (2 900 000 versus 495 000) and in part to the preferential scission of the poly(methyl methacrylate) chain. This point was confirmed by running tests on a mixture of the two homopolymers in the presence of a radical acceptor to prevent macroradical recombination, and on the isolated block copolymers.

Acrylonitrile monomer when masticated in the presence of polymer leads to the formation of pseudocrosslinked block copolymers by mechanical scission of soluble block copolymers. The aggregation of the polyacrylonitrile chains of the block copolymer fraction results in the formation of swollen gels when the polymerization products are extracted with solvents from the initial polymer (78–80).

Ceresa (78, 79) studied in detail the system poly(methyl methacrylate)-acrylonitrile. Figure 25 shows the change in composition with mastication time. A study of gel formation by the block copolymers was made by subjecting the isolated fractions of block copolymers to further mastication. A wide range of block copolymers with varying composition and structure was obtained (Fig. 26).

An interesting study was performed by Guyot and coworkers (86–88) on poly(vinyl chloride) systems using the Brabender Plastograph. They divided the added monomers in two classes: those which possess sufficient "intermolecular"

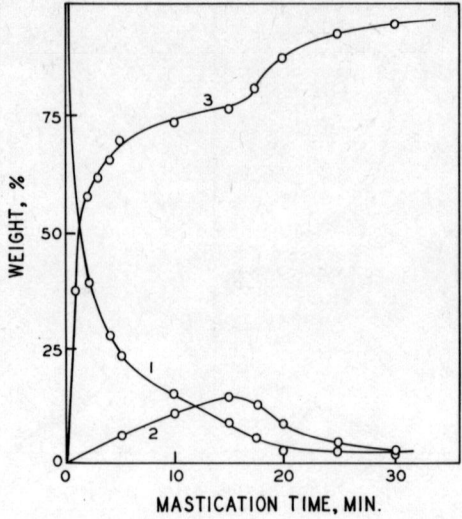

Fig. 25. Polymerization of acrylonitrile by poly(methyl methacrylate) mastication. Analysis of products at various extent of polymerization. *1* free poly(methyl methacrylate); *2* soluble block copolymer; *3* chloroform-insoluble block copolymer gel (78)

plasticizing powers, and monomers which have "interstructural" plasticity ability, following the classification by Kargin of plastifiers. If the interaction between monomer and polymer is greater than polymer-polymer interaction, the monomer with intermolecular plasticizing power can be easily grafted by mechanical synthesis on the poly(vinyl chloride); if the interaction between monomer and polymer is low, the monomer does not readily react with poly(vinyl chloride). Such latter monomers possess an "interstructural" plasticizing power and the mixture monomer-polymer cannot achieve the viscoelastic state necessary for good mechanical scission.

The addition of an intermolecular plasticizer, *i.e.*, a monomer of the first class, allows the graft reaction. Table 18 summarizes these results. The Brabender curves (torque or "consistency" versus time) generally go through a maximum, or so-called "peak", before leveling out. These are the "peak" values given in Table 18.

Guyot and coworkers have also noted that monomers stabilize poly(vinyl chloride) during mastication, probably because they act as free radical scavengers with the exception of basic monomers, such as vinyl pyridine, B. In this case dehydrochlorination reaction

$$\sim\underset{\underset{B}{\overset{}{H}}}{\overset{\overset{Cl}{|}}{CH}}\!-\!CH\sim \longrightarrow \sim CH\!=\!CH\sim + BH^{\oplus} + Cl^{\ominus}$$

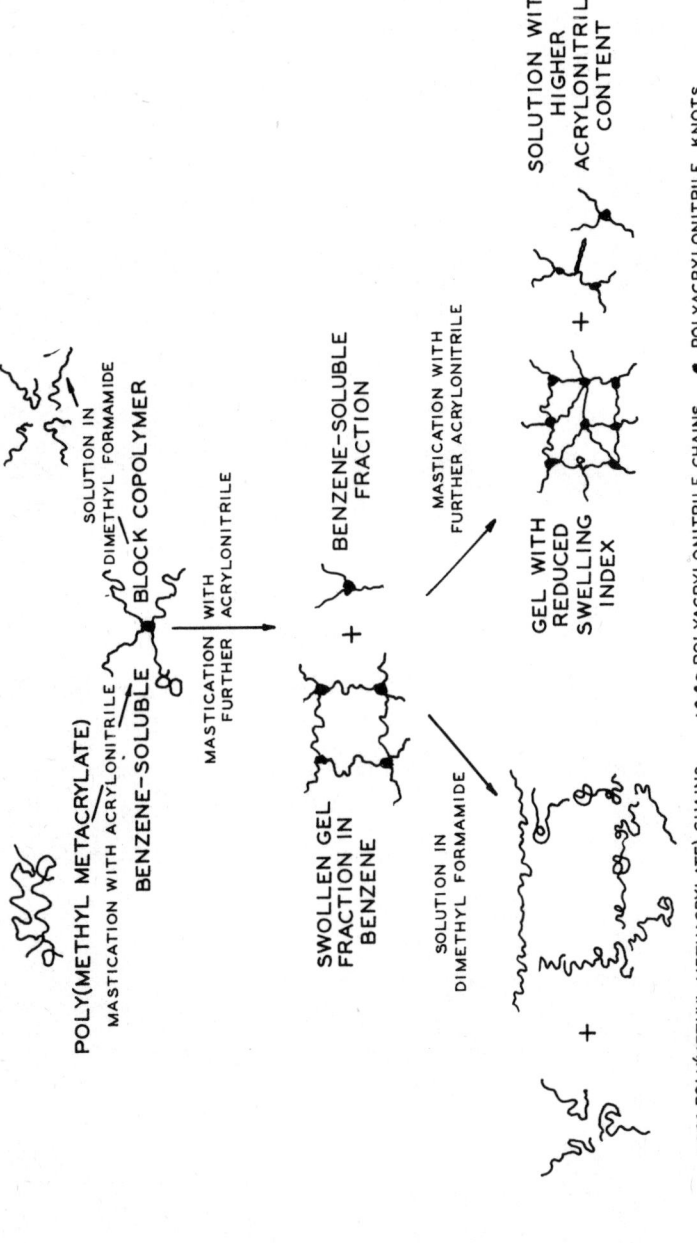

Fig. 26. Polymerization of acrylonitrile by poly(methyl methacrylate) mastication. Schematic representation of block copolymer formation and solvent behavior (78)

Table 18. Polymerization of various monomers by poly(vinyl chloride) mastication (86)

Monomer	Composition				Time min	Gelification time min	Peak Consistance mkp	T, °C	Conversion %	Interpolymer type
	PCV g	Monomer ml	DOP ml	THF ml						
Methyl methacrylate	21	8	0	0	22	5	—	66–98	90	Colorless
Methyl methacrylate	21	6	0	5	10	1	—	58–49–68	90	Colorless
Methyl methacrylate	21	5	5	0	35	2	—	69–80	80	Colorless
n-Butyl methacrylate	20	8	0	0	35	8	—	74–59	81	Milk white
Heptyl methacrylate	22	7	0	0	25	5	—	79–60	78	Milk white
Allyl methacrylate	20	5	6	0	25	9	—	65–70	91	Milk white
Cyclohexyl methacrylate	20	6	5	0	75	43	—	74–59	76	Colorless
Triethyleneglycol methacrylate	22	5	0	5	7	1	—	65–77	100	Colorless
Dimethyl aminoethyl methacrylate	22	7	0	0	23	3	—	82–84	100	Black crosslinked
Methyl acrylate	20	6	5	0	50	2	3.1–3.3	54	2	Colorless
Ethyl acrylate	22	5	5	0	30	1	3.4–3.3	66–68	0	Colorless
Ethyl acrylate	22	7	0	0	23	3	3.8–3.9	72–74	0	Colorless
n-Butyl acrylate	20	6	5	0	100	5	3.1–3.2	60	2	Colorless
Heptyl acrylate	22	7	0	0	30	2	3.4	78–76	0	Slight colored
Heptyl acrylate	22	5	5	0	35	6	3.1–2.9	70–58	12	Milk white
2-Ethylhexyl acrylate	22	7	0	0	35	3	3.5	79–83	19	Green yellow
2-Ethylhexyl acrylate	22	5	5	0	50	21	3.4–3.3	68–58	34	Brown yellow
2-Ethylhexyl acrylate	22	5	0	4	30	1	3.2–3.0	72–78	24	White
Diethyl maleate	22	10	0	0	43	3	3.5–2.9	74–67	8	Brown
Diethyl maleate	22	7	5	0	73	10	2.8	62–50	2	Yellow
Diallyl maleate	22	7	0	0	33	3	3.5	79–72	3	White
Diallyl maleate	22	5	5	0	57	8.3	3.5	66–54	8	Brown yellow
Diallyl fumarate	22	6	5	0	50	9	2.8	60–53	4	Colorless
Diallyl fumarate	22	5	0	5	42	1	3.0	62–78	14	Yellow
2-Vinyl pyridine	20	8	0	0	40	1	3.0	70–71	4	Brown
4-Vinyl pyridine	20	8	0	0	60	1	3.4–3.7	64–79–76	—	Brown
Acrylonitrile	20	6	4	0	50	5	3.2–3.1	60–64	6	Colorless
Methacrylonitrile	20	8	0	0	50	7	3.4–3.3	58–60	2	Colorless
Vinyl acetate	22	5	5	0	35	1	3.3–3.5	68–70	0	Colorless
Vinyl acetate	22	5	0	5	30	2	2.9–	58–62	3	Colorless
Vinyl methyl ketone	22	5	1	0	30	1	4.0	72–81	2	Yellow
N-Vinyl pyrrolidone	22	5	5	0	30	1	3.0	55–50	0	Yellow

prevails over the substitution reaction

$$\sim\text{CH}-\underset{\underset{\ddot{\text{B}}}{|}}{\overset{\overset{\text{Cl}}{|}}{\text{CH}}}\sim \longrightarrow \sim\text{CH}-\underset{\underset{\text{B}^{\oplus}}{|}}{\overset{\overset{\text{H}}{|}}{\text{CH}}}\sim + \text{C-}^{\ominus}$$

for proton affinity.

This technique is illustrated for the case of styrene (interstructural monomer) and methyl methacrylate (intermolecular). Figure 27 shows the plastogram obtained using methyl methacrylate as monomer. The polymerization begins at point B, reaches 40% of conversion at point D, 60% at point E and 100% at F. The interpolymer was characterized by analysis, pyrolysis, fractionation, IR and NMR. It was determined that the interpolymer is composed of three fractions: 70% is poly(vinyl chloride) containing up to 1.4% of methyl methacrylate, 7% is a graft polymer containing 72% of poly(vinyl chloride) and 28% of methyl methacrylate, the third fraction (20% of the total) is poly(methyl methacrylate) with only 4–5% of vinyl chloride chain units. The absence of lactone groups by IR analysis showed that the number of radicals due the scission of poly(vinyl

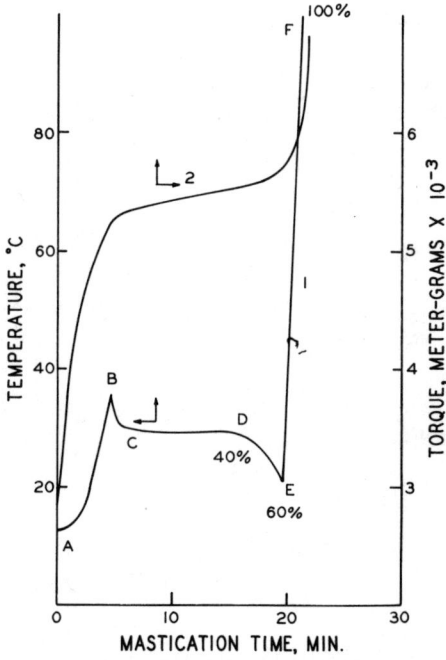

Fig. 27. Plastogram of a mixture of poly(vinyl chloride) (21 g) and methyl methacrylate (intermolecular monomer) (8 ml) (86)

chloride) chains (and thus able to initiate the methyl methacrylate polymerization) is very low. Lactones are formed by the following reactions between adjacent vinyl chloride and methyl methacrylate units:

$$\sim CH_2-CH-CH_2-\underset{\underset{CH_3O-C=O}{|}}{\overset{\overset{CH_3}{|}}{C}} \longrightarrow \sim CH_2-CH-CH_2-\underset{\underset{C=O}{|}}{\overset{\overset{CH_3}{|}}{C}}\sim + CH_3Cl$$

$$\sim CH-CH_2-CH_2-\underset{\underset{CH_3-O-C=O}{|}}{\overset{\overset{CH_3}{|}}{C}} \longrightarrow \sim CH-CH_2-CH_2-\underset{\underset{C=O}{|}}{\overset{\overset{CH_3}{|}}{C}}\sim + CH_3Cl$$

The results suggests that the copolymer has a graft structure and that the mastication medium involves three kinds of domains. The first is the inner domain of poly(vinyl chloride) which is only slightly penetrated by monomer. Polymerization is initiated by macroradicals created in the PVC domain causing the formation of a true copolymer. Short radical segments arising from transfer reactions migrate into the third external domain which consists practically entirely of pure monomer and there initiate polymerization. The second domain is the surface of the resin particle which is swollen by monomer. The free radicals created by bond rupture appear in this second domain.

Figure 28 shows the plastogram for an interstructural monomer (styrene). The mechanochemical synthesis by mastication was also applied to natural polymers (80, 82). The results are reported on Tables 19 and 20.

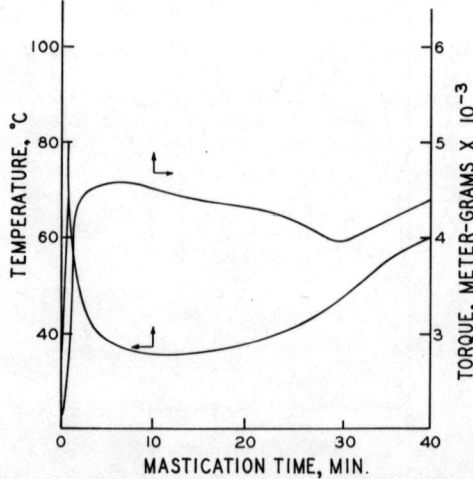

Fig. 28. Plastogram of a mixture of poly(vinyl chloride) (21 g) and styrene interstructural monomer) (8 g) containing 1 % of benzoyl peroxide (86)

Table 19. Polymerization of methyl methacrylate by natural polymer mastication (82)

Polymer	Monomer %	Mastication time min	Monomer conversion %
Ethyl cellulose	38.5	30	87
Benzyl cellulose	35.2	20	97
Cellulose acetate	25.3	20	95
Casein	25.6	50	90
Starch	25.4	20	98
Chlorinated rubber	27.5	20	98
Run congo copal	23.3	35	88
Pale congo copal	15.1	30	89
Mamilla copal	17.4	40	97
Shellac	23.2	20	96
Animal bone glue	15.0	25	83

Table 20. Polymerization of vinyl polymers by mastication of cellulose derivatives (80)

Polymer A	Monomer B	B %	Mastication time min	Conventional[a] precentage	Poly A	Poly B	Block AB
Methyl cellulose	Acrylonitrile	23.7	25	89	29.0	—	71.0
Ethyl cellulose	Methyl methacrylate	38.5	30	93	49.9	4.9	45.2
Benzyl cellulose	Styrene	35.1	20	87	42.9	1.1	56.0
Cellulose acetate	Vinal acetate	25.0	25	95	39.6	14.4	46.0
Starch	Methyl methacrylate	20.2	20	98	35.4	6.0	58.6

[a] Original manuscript (20) suggests that "conventional percentage" corresponds to percent monomer conversion.

3. Molten State

A definition of a boundary between a rubbery and a molten state is inevitably arbitrary. Above T_g amorphous polymers of sufficiently high molecular weight are in a viscoelastic (rubbery) state. Increasing the temperature further, the balance between viscous and elastic contributions changes. Before reaching a purely viscous state, thermal decomposition appears in high polymers. At temperatures required for polymer processing, mechanochemical degradation is accompanied by thermo-oxidative reactions which become even more important with increasing temperature. Processes are considered to be in the molten state if degradation is carried out in standard equipments and under conditions similar to those applied in the industry for conventional processing. The synthesis of block and graft copolymers in the molten state is then the result either of mechanochemical degradation or of thermal decomposition of macromolecules at elevated temperatures. The former increases, as usual, with reduction of

Table 21. Changes in mechanical strength for modified polyethylene as a result of thermal oxidative aging (90)

Material	Tensile strength, kg/cm^2		Strength loss by aging %
	Before aging	After aging	
High density polyethylene	248	170	30
+ 5% SKS-30-1	257	230	12
+ 5% SKS-30-1 + 1% ZnO	310	280	10
+ 5% SKS-30-1 + 1% MgO	295	255	13
+ 10% SKS-30-1	206	170	14

SKS-30-1 = styrene rubber containing methacrylic acid.
[a] Tests performed in an Atlas weatherometer for 100 h.

temperature and with increasing mechanical action. Kerber has prepared a review on the modification of polymers during processing (6).

The modification of homopolymers by production of blocks and grafts by using the lower limit of thermal decomposition has been applied by Akutin and coworkers to the modification of polyethylene with rubber (89–92). They called this process "mechanical-pyrolitic synthesis".

They extruded high density polyethylene without stabilizers and polyisobutylene of molecular weight 203 000 at 200–300° C. The presence of a block and graft copolymer was demonstrated by selective extraction and IR analysis (89). Investigations showed that the melting points of the original and modified polyethylene were the same, but that the thermal stability improved. The incorporation of 5% of polyisobutylene in the process of extrusion only slightly altered the tensile strength and the elongation of the polyethylene.

In a second experiment (90) they selected a butadiene-styrene-carboxylate rubber as a modifying agent for its "compatibility" with polyethylene and its ability to form salt complexes with bivalent metals (ZnO, MgO). The synthesis was performed in a laboratory extruder at 170–210° C. The conversion of polyethylene was studied by X-ray diffraction. The addition of 5% of rubber increased the density of the original polyethylene, with the formation of a more perfect and finely divided crystal structure. An increase in the amounts of modifying agent reduced the crystallinity as determined by NMR. By varying the amount of rubber and of oxide and using the optimum processing temperature, a material can be produced with improved mechanical strength and deformation characteristics. The structural modification of polyethylene improved also the viscous flow properties, the resistance to thermo-oxidative degradation and, most importantly, to aging, see Table 21.

Akutin (52) performed an exhaustive study on the influence of processing conditions and the nitrile rubber/poly(vinyl chloride) ratio on the mechanical characteristic of the product. The experiments were carried out in a Brabender Plastograph at 160–180° C and rotor speed of 10–50 rpm. The poly(vinyl chloride) molecular weight was 83 500. A resin with an epoxy group content of 20.75% was added to the blend as plasticizer and stabilizer. Figure 29 shows that the

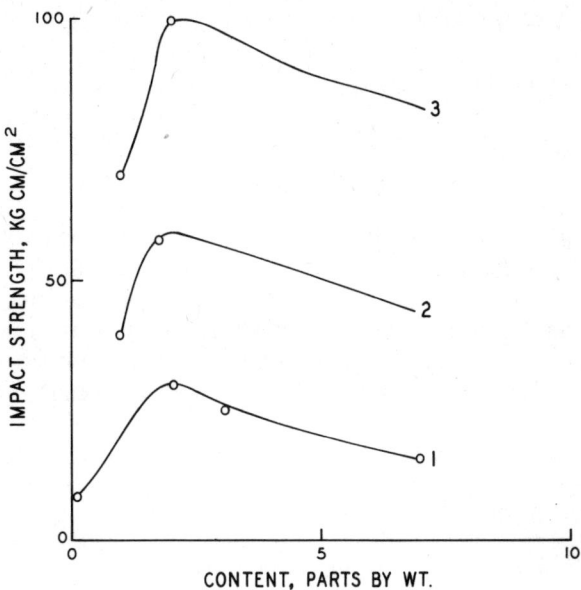

Fig. 29. Dependence of the impact strength of the composition poly(vinyl chloride), epoxy resin, nitrile rubber on rubber content. *1* 0% of SKN 18M; *2* 5% of SKN 18M; *3* 10% of SKN 18M *(52)*

addition of nitrile rubber sharply increases the impact strength of the composition. The addition of rubber up to 10% leads to a reduction in torque, which indicates a plasticizing action of the rubber. By raising the content of nitrile rubber to 50% the torque increases sharply due to the formation of gel. It may be noted that this torque increase was not found in the case of pure rubber and for a blend of poly(vinyl chloride) with an incompatible styrene rubber. By THF extraction the 10 and 50% nitrile rubber blends gave respectively 7 and 12% of an insoluble fraction. Elementary analysis on the fractions showed the presence of chlorine and nitrogen.

Baramboim and coworkers *(93–95, 95 bis)* obtained graft copolymers based on polyamides. The extrusion of a polypropylene polycaprolactam mixture at 200–210° C changed the polypropylene molecular weight and formed a block and graft copolymer of linear and three dimensional structure. Investigations showed the radical nature of the process and that the amount of the resulting copolymer was proportional to the intensity of mechanical shear.

The possibility of modifying polypropylene by extrusion at 220° C in the presence of alkaline sulphate lignin and a plasticizer to increase the low temperature resistance was also studied *(96)*. The modifying effect of the lignin appears to be due to its rupture into free radicals which may cause crosslinking of the polypropylene macromolecules. The polypropylene which was modified (Propolin) is used to make films which have increased low temperature and light resistance.

Mastication of extracted natural rubber with methyl methacrylate in an extruder gives a relatively high conversion of monomers, >60%, and the formation of a modified rubber (97).

4. Solution Systems

The mechanochemical synthesis of graft and block copolymers in solution will likely be not so important as in the undiluted state. This is because the conversion of mechanical energy to free radical generation is much less efficient. A popular degradation method for solutions is by ultrasonic irradiation. It is a method of high power input. A serious disadvantage is the energy inhomogeneity. There is also the limitation of combining only polymers and monomers soluble in the same solvent.

Solution synthesis can be conducted by some irradiation and by capillary shear. The former is unfortunately not covered here, the latter is reported only in the Brit. pat 679,562.

a) High Speed Stirring

If a mechanical degradation of a solution of two polymers is carried out by high speed stirring, the formation of a block copolymer is not probable as the scission of polymer molecules at low concentration is not caused mainly by intermolecular interaction, such as by collision of molecules and through entanglements, but by displacements due to hydrodynamic forces in velocity gradients. Nakamo and Minoura (98) did obtain reaction by stirring a benzene solution of polyethylene oxide and poly(methyl methacrylate).

If mechanical degradation of a polymer solution by high speed stirring is carried out with a vinyl monomer in an inert atmosphere the monomer will be polymerized by the free macroradical. Goto and Fujiwara, for example, studied poly(vinyl acetate vinyl acetate) agitated in an Homomixer at 30000 rpm in a nitrogen atmosphere at 65° C (99). They found that the polymerization rate, V_p, is proportional to the square root of the initial concentration of poly(vinyl acetate):

$$V_p = (2k_d/k_t)^{1/2} k_p (M)(P_0)^{1/2}$$

where k_d, k_t, k_p, are respectively the degradation, termination and propagation rate constants, (M) and (P_0) the monomer and initial polymer concentration. No polymerization occurred when hydroquinone was added or when the polymer concentration was lower than 6.88 g/l. The limiting chain length of degradable poly(vinyl acetate) also increased as the concentration of the solution decreased.

A detailed study on the possibility of carrying out block copolymerizations by high speed stirring was performed by Minoura and coworkers (98, 100) on the system polyethylene oxide-methyl methacrylate. The polymer was dissolved in the monomer at 45° C and stirred in high speed homomixer in a nitrogen atmosphere. The polymer molecular weight was in the range of 1000000, the speed of stirring near 30000 rpm and polymer concentration generally 4 g/100 ml. They investigated the following parameters: monomer and polymer concentration,

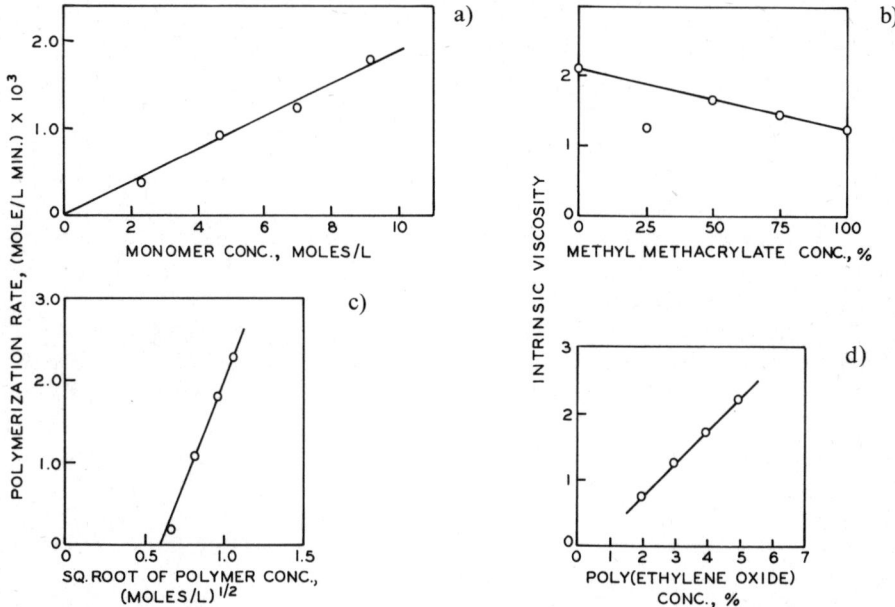

Fig. 30a–d. Polymerization of methyl methacrylate by high speed stirring of polyethylene oxide solution. a) effect of monomer concentration on polymerization rate (PEO 4 g/100 ml, stirring speed 30000 rpm. b) effect of monomer (MMA) concentration on intrinsic viscosity of reaction mixture (PEO 4 g/100 ml, stirring speed 30000 rpm, solvent: benzene. c) effect of PEO concentration on polymerization rate. d) effect of PEO concentration on intrinsic viscosity of reaction mixture (Stirring speed 30000 rpm)

polymer molecular weight, stirring speed, and solvent effect. In all cases an almost constant rate of conversion was obtained. The rate was proportional to monomer concentration (Fig. 30a), while the intrinsic viscosity of the mixture decreased with increasing concentration (Fig. 30b). By comparing the results with those on stirring a solvent solution of polyethylene oxide (101), it was concluded that the effect of all other parameters is connected with the rate of macroradical formation. A relationship was also found between the polymerization rate and the square root of polyethylene oxide concentration, in agreement with Goto (Fig. 30c). The intrinsic viscosity of the reaction mixture increased linearly with polyethylene oxide concentration (Fig. 30d). The higher the molecular weight, > 38000, the more monomer polymerized. The intrinsic viscosity of the resulting polymer decreased with time. Increasing the speed from 10000 to 30000 increased the amount of methyl methacrylate polymerized. The polymerization rate was also faster in systems containing solvents where polyethylene oxide was degraded more efficiently.

By the above results and by the general rate equations for initiation, propagation, and termination, Minoura concluded that the polymerization rate is proportional to the monomer concentration (M) and to the square root of the difference between the total polymer concentration (P) and the concentration required for the initiation of polymerization (P_α):

$$V_p = k_p (f k_d / k_t)^{1/2} (M) [(P) - (P_\alpha)]^{1/2}$$

where k_p, k_d, and k_t are the propagation initiation and termination rate constant and f is the proportion of the radicals concentration effective for initiating the propagation of monomers.

The radical reaction mechanism was confirmed by polymerizing a mixture of styrene and methyl methacrylate. The ratio of the monomers in the copolymer (1.15) was nearly equal to the value (1.05) calculated from the reactivity ratio for radical copolymerization and differed considerably from the value of 10.5 for the cationic copolymerization and from the value 0.15 for anionic copolymerization (78).

b) Freezing and Thawing

Freezing and thawing of aqueous solutions of macromolecules can lead to formation of free macroradicals and to degradation. The reaction is due to an increase of specific volume and the consequent development of mechanical stresses. Berlin and coworkers (102, 103) were the first to apply this technique by subjecting to repeated freezing an emulsion of polystyrene with toluene in 2.5–5% starch solution. The product separated after the coagulation of the thawed emulsion gave opalescent solutions in water and toluene, indicating the formation of a copolymer containing high polymer blocks of starch and low molecular weight blocks of polystyrene.

Ceresa (80) demonstrated the possibility of synthetizing block copolymer by subjecting a starch emulsion with free radical polymerizable monomers to repeated freezing at $-200°$ C and subsequent thawing to room temperature. He used acrylonitrile owing to the case of separating the insoluble block copolymer fraction, see Table 22.

c) Vapor Phase Swelling

If a polymer is allowed to absorb solvents from the vapor phase it will swell until equilibrium is reached. A critical degree of swelling exists where the rate of diffusion changes suddenly. This critical swelling depends on the nature and the molecular weight of the polymer and the solvent employed. At the critical swelling, the rupture of the weaker linkages of the chain occurs. If a polymerizable solvent monomer is used, it can block copolymerize in the swollen polymer by the macroradicals produced in situ. The existence of the critical degree of swelling in a system polymer-monomer was demonstrated by Ceresa for the cellulose acetate-acrylonitrile vapor system (80). The change in swelling rate occurred

Table 22. Polymerization of acrylonitrile by freezing and thawing starch emulsion (80)

Starch: acrylonitrile in 10% emulsion	Freezing cycles, number	Block copolymerization %
1:1	5	3.5
	10	6.8
	15	8.2
	20	13.4
	35	26.8
1:3	5	4.9
	10	11.2
	15	13.7
	20	21.7
	25	38.9
3:1	5	2.4
	10	8.9
	15	11.7
	20	26.3
	25	27.4

The emulsion was stabilized with 0.25% block copolymer of starch and methyl methacrylate synthesized by mastication.

Table 23. Polymerization of acrylonitrile by swelling of cellulose acetate vapor (80)

Swelling min	Monomer %	Polymerization %	Product composition	
			Cellulose acetate %	Block copolymer %
10	6.8	nil	100	—
20	10.9	nil	100	—
30	14.6	nil	100	—
40	15.1	1.6	86.2	13.8
50	19.4	3.8	80.6	19.4
60	24.6	13.4	64.1	35.9

after 40 min swelling at a monomer uptake of 15%. Below this point no polymerization occurred, see Table 23.

Ceresa synthetized also block copolymers of poly(methyl methacrylate) with acrylonitrile and styrene and of polyethylene with methyl methacrylate, styrene using this method (104).

More recently, possibility has been demonstrated of grafting in yield a wool-aqueous ethyl acrylate system. The grafting solution needs to be left in contact with the sample for long time periods (105). After an induction period, the reaction rate is very fast, see Fig. 31. It was believed that grafting is initiated by the radical formed initially by anisotropic swelling with water. As grafting proceeds, the monomer becomes the predominantly swelling agent and this coupled with the

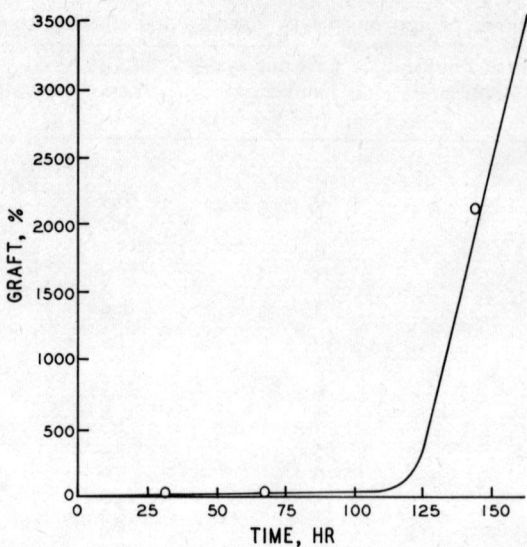

Fig. 31. Polymerization of ethyl acrylate by wool swelling at 25° C using water as a swelling agent (60% aquesous emulsion with 3.2% Triton X-405) (*105*)

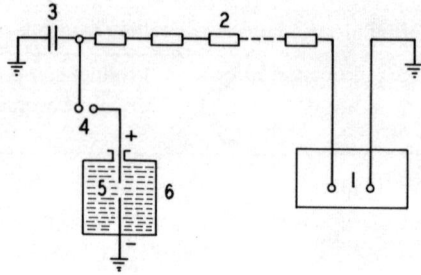

Fig. 32. Spark-discharge circuit: *1* H.V. rectifier giving recified voltages in the range 50–100 kV; *2* Charging resistors; *3* H.V. pulse capacitor, 0.56 µF; *4* Activating spark-gap; *5* Main spark-gap; *6* Liquid

gel effect presumably results in the higher subsequent grafting rates. The formation of macroradicals was demonstrated by ESR. The influence of temperature was investigated using regenerated cellulose fibers. Increasing the temperature, the induction time decreased with an activation energy of 52.3 kcal/mole.

d) Spark Discharge

This system of synthesis was studied by Akutin and coworkers (*4, 106–108*). A spark gap, *4*, see Fig. 32, allows a sudden release at the main spark gap, *5*, the energy accumulated in a capacitor, 3, at the moment of the spark discharge in the liquid, 6. This discharge takes only microseconds and it is accompanied by a peak of current with an amplitude of some thousand amperes. At the same time, a sharp rise of pressure takes place in the discharge channel, which for the low compressibility of the liquid is conveyed in the form of hydraulic shock. The liquid forms voids on moving in a radical direction from the spark zone. The subsequent dissipation results in a cavitational shock, the force of which is added to that of the main hydraulic discharge shock. To produce graft and block copolymerization, the spark discharge was applied at a frequency of 0.5–1.5 c/s. The pressure impulses produced macromolecular rupture, radical formation and reaction. In presence of a monomer, block copolymers were synthetized. Solutions of poly(vinyl chloride) or of poly(trifluoroethylene) in methyl methacrylate were used.

5. Polymer-Filler Interactions

Polymer may react with filler, or other nonpolymerizable materials, through mechanochemical reaction, converting them into an insoluble three dimensional gel. Reactions leading to reinforcement of rubber are a typical example, largely employed on an industrial scale. The more important of these reactions are the combination of rubber and carbon black or aluminium alkoxides. The structure of carbon black likely provides sites for attack of radicals, indeed carbon black is a radical acceptor of a special polyfunctional type, as a particle may terminate more than one sheared rubber chain. The chemical nature of the carbon rubber interaction through a mechanochemical reaction has been demonstrated by Watson (109) in an early work. The strongest evidence for chemical combination by radical formation was the negative temperature coefficient, the influence on gel content by radical acceptors, the different behavior of natural and synthetic rubber — in good agreement with their cold milling behavior. The reader is referred to the specialized book, as regards details on the reinforcing effect of carbon black.

Ceresa (110) investigated the cold mastication of natural rubber with compounds of the general form MX_n, where M is a metal with a valency greater than 2 and X is a halide, hydroxide or nitrate. The reaction leads to the rapid formation of an insoluble gel fraction. The most effective gelling agents were found to be halides of aluminum, iron and tin, but their hygroscopicity raises a question. It was thus more convenient to employ aluminium alkoxides. Their efficiency is an inverse function of alkyl chain length. The addition of aluminum isopropoxide aids the formation of block copolymer. This technique was applied to the system polyethylene-poly(vinyl acetate). Experiments were run with natural and synthetic rubber by using an internal mixer, roll mills and an extruder. As the gelation is a mechanochemical process, independent of the elastomer structure, the reaction was applied to plastomers masticated while in the viscoelastic state (111). The reaction takes place with poly(methyl methacrylate), poly(vinyl acetate) and polyethylene. No evidence for crosslinking of polystyrene was observed.

Ceresa (112) used the mechanical synthesis of natural rubber with phenolic resins to achieve rubber reinforcement by the generation of gel. The reaction was performed in the Baker-Perkins Unirotor masticator at 100 rpm, 12° C, and in a nitrogen atmosphere. Eight different commercially available phenol-formaldehyde resins were used after grinding in a vibromill. The rates of gelation were all the same order for acetone extracted rubber masticated with 1% of each phenolic resins. With the most reactive resin a maximum gel content was achieved after 40 min; whereas, the other resins required 60–150 min. The maximal gel content varied from 73 to 85% and did not change significantly during mastication up to three hours. An increase in the concentration of resin resulted in a small but consistent increase in gelation rate. Unextracted rubber or oxygen lowered the gel content. Active acceptors inhibited the reaction. The order of resin reactivity remained essentially the same under different mastication conditions. A decrease in shear rate by either changing scroll design or reducing rotor speed or raising the scroll temperature or plasticizing with solvent decreased the gel content. A reaction between the active chains produced by mechanical scission of rubber and the reactive groups attached to the resin was postulated, since the behavior above is typical of a mechanical reaction. The complete dispersion of resins on a molecular scale, as recorded by electron micrographs, confirmed that interactions of rubber with resin were chemical in nature. Therefore, a series of mastications of natural rubber with phenols with functionalities from 1 to 6 were carried out to evaluate relative reactivity. A functionality of three, of course, is required for gelation to occur (*112*).

Acknowledgement

The authors wish to express their sincere appreciation for the most helpful suggestions made by Professor John D. Ferry, Editor for Advances in Polymer Science.

E. References

1. Ceresa, R. J.: Block and Graft Copolymers, Ch. 5. London: Butterworths 1962.
2. Batterd, H. G., Tregear, G. W.: Graft Copolymers. New York: Interscience Publ. 1967.
3. Allport, D. C., Janes, W. H.: Block Copolymers. London: Applied Science Publ. 1973.
4. Akutin, M. K.: Plast. Inst. Trans. **28**, (78) 216 (1960).
5. Berlin, A. A.: Usp. Khim. **27**, 94 (1958); CA **52**/11456b.
6. Kerber, M. L.: Soviet Plastics (5) 64 (1971).
7. Ceresa, R. J.: J. Polymer Sci. **53**, 9 (1961).
8. Vasiliu-Oprea. C., Simionescu, C.: Materiale Plast. **3**, (2) 64 (1966).
9. Minoura, Y., Kawamura, S.: Yuki Gosei Kagaku Kyokai Shi **23**, (5) 394 (1965).
10. Melville, H. W.: Plastics Progr. 1 (1955).
11. Baramboim, N. K.: Mechanochemistry of Polymers, translated from Russian by R. J. Moseley, edited by W. F. Watson, Rubber and Plastic Research Association of Great Britain, Maclaren, 1964.
12. Casale, A., Porter, R. S., Johnson, J. F.: Rubber Chem. Tech. **44** (2) 534 (1971).
13. Simionescu, C., Vasiliu-Oprea, C.: Mechanochimia Compusilor Macromoleculair, Ed. Academiei Republicii Socialiste Romania, 1967.
14. Deters, W., Huang DeHja-Chian: Faserforsch. Textiltech. **14**, 58 (1963).
15. Baramboim, N. K.: Zh. Fiz. Khim. **32**, 806 (1958).
16. Bischof, K.: Rev. Chem. (Bucharest) **13**, 205 (1962); CA **57**, 16862i.
17. Whistler, R. L., Goatley, J. L.: J. Polymer Sci. **62**, S123 (1962).
18. Grohn, H., Vasiliu-Oprea, C.: Rev. Roumaine Chim. **11** (11), 1297 (1966).
19. Groh, H., Bischof, K.: Plaste Kautschuk **8**, (6), 311 (1961).
20. Bischof, K., Korn, R.: Plaste Kautschuk **10**, 28 (1963); CA **59**, 6532f.
21. Bischof, K., Korn, R.: Plaste Kautschuk **10**, 80 (1963); CA **59**, 10254f.
22. Dubinskaya, A. M., Butyagin, P. Yu., Berlin, A. A.: Dokl. Akad. Nauk SSSR. **159** (3) 595 (1964).
23. Simakov, Yu. S., Baramboim, N. K.: Nauchn. Tr. Mosk. Tekhnol. Inst. Legkoi Prom. **36**, 59 (1969).
24. Simakov, Yu. S., Baramboim, N. K.: Nauchn. Tr. Mosk. Tekhnol. Inst. Legkoi Prom. **35**, 102 (1969).
25. Hsu, H., Chang, S., Wu, H., Hsu, A., Li, C., Hsu, T., Lo, C., Chou, S., Hsueh, K.: Chu Pan She **1963**, 369.
26. Protasov, V. G., Baramboim, N. K.: Plast. Massy (2) 8 (1969). Soviet Plastics **2**, 5 (1969).
27. Grohn, H., Bischof, K., Hensinger, H.: Plaste Kautschuk **9**, 180 (1962).
28. Crusos, A., Feldman, D., Simionescu, C.: Rev. Roumaine Chim. **13**, 1489 (1968).
29. Baramboim, N. K., Simakov, Yu. S.: Polymer Sci., U.S.S.R. (Engl.) **8**, 235 (1966).
30. Baramboim, N. K., Simakov, Yu. S.: Nauchn. Tr. Mosk. Tekhnol. Inst Legkoi Prom. **30**, 188 (1964).
31. Grohn, H., Vasiliu-Oprea, C.: Analele Univ. Bucuresti Ser. Stiint Nat. **14**, (1), 27 (1965); CA **65**, 20284b.
32. Grohn, H., Vasiliu-Oprea, C.: Rev. Roumaine Chim. **9**, 757 (1969).
33. Vasiliu-Oprea, C., Neguleanu, C., Simionescu, C.: E. Polymer J. **6**, 181 (1970).
34. Vasiliu-Oprea, C., Neguleanu, C., Simionescu, C.: Plaste Kautschuk **17**, (9) 639 (1970).
35. Vasiliu-Oprea, C., Neguleanu, C., Simionescu, C.: Makromol. Chem. **126**, 217 (1969).
36. Simionescu, C., Vasiliu-Oprea, C., Neguleanu, C.: Makromol. Chem. **148**, 155 (1971).
37. Simionescu, C., Vasiliu-Oprea, C., Neguleanu, C.: Bul. Inst. Politeh Iasi **15**, (3–4) 45, (1969).
38. Grohn, H., Bischof, K.: Chem. Tech. **11**, 384 (1959).
39. Todd, A.: J. Polymer Sci. **42**, 223 (1960).
40. Simionescu, C., Feldman, D., Crusos, A.: Bull. Soc. Chim. **4**, 1525 (1969).
41. Grohn, H., Vasiliu-Oprea, C.: Plaste Kautschuk **13**, 385 (1966).
42. Baramboim, N. K., Komissarov, S. A., Simakov, Yu. S.: Nauchn. Tr. Mosk. Tekhnol. Inst. Legkoi Prom. **35**, 93 (1966).
43. Watson, W. F., Wilson, D.: J. Sci. Instr. **31**, 98 (1954).

44. Angier, D. J., Watson, W. F.: J. Polymer Sci. **18**, 129 (1955).
45. Angier, D. J., Watson, W. F.: Trans. I.R.I. **33**, 22 (1957).
46. Dogadkin, B. A., Kuleznev, V. N., Tarasova, Z. N.: Kolloidn. Zh. **20**, 43 (1958).
47. Dogadkin, B. A., Zuleznev, V. N.: Kolloidn. Zh. **20**, 674 (1958).
48. Staudner, E., Beniska, J.: Chem. Zvest. **16**, 431 (1962).
49. Slinimskii, G. L., Reztsove, E. V.: Vysokomolekul. Soedin **1** (4), 534 (1959); — Rubber Chem. Tech. 457 (1960).
50. Kargin, V. A., Kovarskaya, B. M., Golubenkova, L. I., Akutin, M. S., Slonimskii, G. L.: Khim. Prom. **1957**, 77.
51. Kargin, V. A., Plate, N. A., Dobrynina, A. S.: Kolloidn. Zh. **20**, 332 (1958).
52. Akutin, M. S., Melik-Kasumov, A. V., Torneu, R. V., Kotrelev, V. N.: Plast. Massy **12**, 45 (1968).
53. Kovarskaya, B. M., Golubenkova, L. M., Akutin, M. S., Levantouskaya, I. I.: Vysokomolekul. Soedin. **1**, 1042 (1959); — Rubber Chem. Tech. 964 (1960).
54. Kargin, V. A., Kovarskaia, N. M., Golubenkova, L. I.: Dokl. Akad. Nauk. SSSR **112**, 485 (1957).
55. Beniska, J., Staudner, E.: Chem. Zvest. **17**, 330 (1963).
56. Berlin, A. A., Gilman, I. M.: Kauchuk i Rezina **19**, No. 12, 1 (1960).
57. Berlin, A. A., Kronman, A. G., Ianovski, D. M., Kargin, V. A.: Khim. Prom. **2**, 96 (1962).
58. Berlin, A. A., Kronman, A. G., Yanovskii, D. M., Kargin, J. A.: Vysokomolekul. Soedin. **2**, 1188 (1960); — Rubber Chem. Tech. 760 (1961).
59. Kargin, V. A., Plate, N. A., Rebinder, E. P.: Vysokomolekul. Soedin. **1**, 1546 (1959).
60. Shaw, E., Wake, W. C.: R.A.B.R.M. Res. Mem. N. R. **1958**, 412.
61. Fujii, H., Kira, T., Koide, M., Goto, K.: Nippon Gomu Kyokaishi **35**, 28 (1962).
62. Le Bras, J.: Rev. Gen. Caoutchouc **19** (2), 43 (1942).
63. Compagnon, P., Bonnet, O.: Rev. Gen. Caoutchouc **19** (3), 79 (1942).
64. Le Bras, J., Compagnon, P.: Bull. Soc. Chim. **11**, 553 (1944).
65. Le Bras, J., Compagnon, P., Delalande, A.: Rev. Gen. Caoutchouc **33**, 148 (1956).
66. Le Bras, J.: Rev. Gen. Caoutchouc **33**, 149.
67. Le Bras, J., Compagnon, P., Delalande, A.: Compt. Rend. **241**, 61 (1955).
68. Angier, D. J., Watson, W. F.: J. Polymer Sci. **25**, 1 (1957).
69. Angier, D. J., Watson, W. F.: J. Polymer Sci. **20**, 235 (1956).
70. Ceresa, R. J., Watson, W. F.: Trans. I.R.I. **35**, 19 (1959).
71. Baramboim, N. K., Santin, B. V.: Vysokomolekul. Soedin. **2**, 1196 (1960).
72. Pone, D. Z., Rekena, Z., Jirgens, L., Zarane, S.: Uch. Zap., Rizh. Politekh. Inst. **27** (10), 36 (1967).
73. Baramboim, N. K., Popov, V. I.: Nauchn. Tr. Mosk. Tekhnol. Inst. Legkoi Prom. **19**, 54 (1961).
74. Baramboim, N., Sautun, B. V.: Vysokomolekul. Soedin. **2** (8), 1196 (1960).
75. Slonimskii, G. L., Kargin, V. A., Reztsova, E. V.: J. Phys. Chem. Moscow **33**, 988 (1959).
76. Reid, W. S.: J. Soc. Chem. Ind. (London) **68**, 44 (1949).
77. Angier, D. J., Ceresa, R. J., Watson, W. F.: J. Polymer Sci. **34**, 699 (1959).
78. Ceresa, R. J.: Polymer **1**, 477 (1960).
79. Ceresa, R. J.: Polymer (4), 488 (1960).
80. Ceresa, R. J.: Polymer **2**, 213 (1961).
81. Ceresa, R. J.: N.R.P.R.A. Mon. Rep. (1955).
82. Ceresa, R. J.: Thesis, University of London, 1958.
83. Romanov, A., Magarik, S. Ya., Lazar, M.: Vysokomolekul. Soedin. B**9** (4), 292 (1967).
84. Protasov, V. G., Makhmudbekova, N. L., Baramboim, N. K.: Nauchn. Tr. Mosk. Tekhnol. Inst. Legkoi Prom. **36**, 63 (1969).
85. Baramboim, N. K.: Nauchn. Tr. Mosk. Tekhnol. Inst. Legkoi Prom. **25**, 147 (1963).
86. Guyot, A., Michel, A.: J. Appl. Polymer Sci. **13**, 911 (1969).
87. Michel, A., Galin, M., Guyot, A.: J. Appl. Polymer Sci. **13**, 929 (1969).
88. Michel, A., Bert, M., Guyot, A.: J. Appl. Polymer Sci. **13**, 945 (1969).
89. Akutin, M. S., Artememko, B. N.: Plast. Massy **9**, 64 (1967); — Soviet Plastics **9**, 66 (1968).

90. Akutin, M. S., Artemenko, B. N.: Vysokomolekul. Soedin. A **10** (3), 561 (1968); — Polymer Sci., U.S.S.R. **10**, 653 (1968).
91. Akutin, M. S., Artemenko, B. N.: Synthesis Modification and Processing of Polyolefins, Baku, 1967, p. 59.
92. Akutin, M. S.: Vysokomolekul. Soedin. Ser. A **11**, 675 (1969).
93. Rakityanskii, V. F., Baramboim, N. K.: Nauchn. Tr. Mosk. Tekhnol. Inst. Legkoi Prom. **36**, 51 (1969).
94. Takityanskii, V. F., Baramboim, N. K.: Nauchn. Tr. Mosk. Tekhnol. Inst. Legkoi Prom. **35**, 107 (1969).
95. Baramboim, N. K., Rakityanskii, V. F.: Synthesis Modification and Processing of Polyolefins. Baku, 1967, p. 93.
95. bis Baramboim, N. K., Takityanskii, V. F.: Plast. Massy **11**, 34 (1971).
96. Lyvbeshkina, E. G.: Synthesis Modification and Processing of Polyolefins, Baku, 1967, p. 52.
97. Beniska, J., Staudner, E.: Sb. Prac. Chem. Fak. SVST 147 (1961).
98. Nakano, A., Minoura, Y.: J. Appl. Polymer Sci. **15**, 927 (1971).
99. Goto, K., Fujiwara, H.: J. Polymer Sci. B **1**, 505 (1963).
100. Minoura, Y., Kasuya, T., Kawamura, S., Nakano, A.: J. Polymer Sci. A-1, **5**, 43 (1967).
101. Minoura, Y., Kasuya, T., Kawamura, S., Nakano, A.: J. Polymer Sci. A-2, **5**, 125 (1967).
102. Berlin, A. A.: 9th Conf. on General Questions on the Chem. and Phys. of High Molecular Weight Comp., lzd. Akad. Nauk. S.S.S.R., 1956, p. 76.
103. Berlin, A. A., Penskaya, E. A., Volkova, G. I.: Mezh. Simp. Makromol. Khim. Doklady, Moscow, 1960, Sektsiya 3, 334.
104. Ceresa, R. J.: 10th Canadian High Polymer Forum, Montreal, 1960.
105. Williams, J. L., Stannett, V.: J. Polymer Sci. B **8**, 711 (1970).
106. Akutin, M. S., Parlashekevitch, N. Y., Kogan, I. N., Rubynshteii, V. V., Grybkova, R. M.: Plast. Massy **6** (1960).
107. Akutin, M. S., Parlashekevitch, N. Y., Menes, L. I., Rubynshteii, V. V., Kotrelev, V. N.: U.S.S.R. Patent 630,097.
108. Bliznakow, C.: Z. Phys. Chem. **209**, 372 (1958).
109. Watson, W. F.: Ind. Eng. Ghem. **47**, 1281 (1955).
110. Ceresa, R. J.: Int. Rubber. Conf., Washington, Nov., 1959.
111. Ceresa, R. J.: Polymer **1**, 72 (1960).
112. Ceresa, R. J.: Trans. I.R.I. **36**, 211 (1960).

Received April 22, 1974

Polymerization through the Carbon-Sulfur Double Bond

WILLIAM H. SHARKEY

Central Research Department, E. I. du Pont de Nemours and Company, Wilmington, Delaware, USA

Table of Contents

I. Introduction	74
II. Thioformaldehyde	74
A. Introduction	74
B. Preparation	75
C. Toxicity	79
D. Crystal Structure	79
E. Applications	80
III. Thioacetone	80
A. Introduction	80
B. Preparation	81
C. Thioketo and Thioenol Tautomers	82
D. Polymerization	82
IV. Higher Thiocarbonyl Compounds	84
A. Introduction	84
B. Thioacetophenone	84
C. 1-Thioacylaziridines	85
V. Fluorothiocarbonyl Compounds	86
A. Introduction	86
B. Synthesis of Monomers	87
1. Thiocarbonyl Fluoride	87
2. Fluorothioacyl Halides	89
3. Fluorothioketones	90
C. Fluorothiocarbonyl Polymers by Ionic Initiation	92
1. Polythiocarbonyl Fluoride	92
2. Copolymers of Thiocarbonyl Fluoride	96
3. Polymerization of Fluorothioacid Fluorides and Fluorothioketones	97
D. Fluorothiocarbonyl Polymers by Free-Radical Initiation	98
1. Thiocarbonyl Chlorofluoride	98
2. Thiocarbonyl Fluoride	99
VI. References	102

I. Introduction

It is very well known that polymers of high commercial value are obtained from formaldehyde by addition polymerization of its carbon-oxygen double bond. Not so well known is the addition polymerization capability of the carbon-sulfur double bond, probably because none of the polymers so obtained has yet become commercially acceptable. However, the polymerization chemistry of the carbon-sulfur double bond has been the subject of a number of studies and these have defined the preparation and properties of polythioformaldehyde, polythioacetone, polymers from a small number of higher thioketones, and polymers from fluorine analogs of thioaldehydes and thioketones. The monomers have great reactivity beyond polymerization, and their general chemistry has been discussed in earlier reviews (1, 2).

Not surprisingly, thioformaldehyde polymerizes more avidly than any other carbon-sulfur double bond compound. At first the polymer appeared promising because of its high melting point and highly crystalline character. The promise has not been realized because the polymer undergoes degradation in the molten form.

Higher thioaldehydes seem to have been overlooked. Thioketones, however, have received much attention. Work in this area has involved principally thioacetone. This compound polymerizes spontaneously and rapidly, though it is not as active as thioformaldehyde. Some work has also appeared on higher thioketones.

Considerable attention has been given to fluorine analogs. Replacement of hydrogen with fluorine in thiocarbonyl polymers leads to a reduction in melting point and glass transition temperature. Consequently, polymers of these compounds are elastomers. Of these, poly(thiocarbonyl fluoride) is one of the most resilient materials known. Fluorine-containing thiocarbonyl polymers are also stronger and tougher than their hydrogen counterparts.

Inasmuch as there are such great differences between hydrogen and fluorine analogs, these classes will be discussed separately. Since hydrogen compounds are generally considered basic by most organic chemists, these will be discussed first, starting with thioformaldehyde.

II. Thioformaldehyde

A. Introduction

Thioformaldehyde is unknown in the free state, but polymers of it have been obtained by ring opening of trithiane, by removal of hydrogen sulfide from methanedithiol, by reaction of formaldehyde and hydrogen sulfide or sodium sulfide, and by reaction of sodium hydrosulfide with methylene chloride. The earliest report of a product believed to be a polymer of thioformaldehyde was that by Wohl (3) in 1886. He obtained a sulfur-containing polymer by reaction of hexamethylenetetramine and hydrogen sulfide. Later (4) it was found that identification of this product as poly(thioformaldehyde) probably was not correct because it contains considerable quantities of nitrogen.

B. Preparation

Authentic poly(thioformaldehyde) was prepared much later by Harmon (5) by removal of hydrogen sulfide from methanedithiol and from bis(mercaptomethyl)sulfide.

$$HSCH_2SH \xrightarrow{R_3N} +CH_2-S+_n + H_2S$$

$$HSCH_2SCH_2SH \xrightarrow{R_3N} +CH_2-S+_n + H_2S .$$

Harmon's method involves treating methanedithiol or bis(mercaptomethyl)sulfide with ammonia, an amine, or a phosphine. When one of these reagents is added to methanedithiol and the mixture heated to slightly over 100° C, hydrogen sulfide is evolved and eventually the reaction mixture becomes solid. The preparation is completed by gradually increasing the temperature of the mixture to 230° C over 3 h and then removing low-molecular-weight materials by extraction with benzene. The product, when most highly purified, melts at 248° C and gives an X-ray diffraction pattern with high intensity lines at 4.43, 3.015, and 2.17 Å.

The methanedithiol polymer precursor and bis(mercaptomethyl)sulfide are formed by reaction of aqueous formaldehyde buffered with sodium dihydrogen phosphate with hydrogen sulfide at 45° C under autogenous pressure (6). Methanedithiol, which boils at 40–44° C/40 mm, is separated from bis(mercaptomethyl)sulfide, bp 65° C/10 mm, by fractional distillation.

Most of the research on thioformaldehyde derivatives has been done on other products formed from formaldehyde and hydrogen sulfide. This reaction is quite complex and gives in addition to methanedithiol and bis(mercaptomethyl)sulfide, trithiane, thioformaldehyde oligomers, and poly(thioformaldehyde).

In 1868 Hofmann (7) reported the preparation of trithiane. This is the product obtained when the hydrogen sulfide-formaldehyde reaction is run under strongly acidic conditions. Under weakly acidic or mildly alkaline conditions, the product formed retains significant amounts of oxygen. Recently, Credali and Russo (8) have examined in depth the reaction of hydrogen sulfide with aqueous formaldehyde free of added acid or base.

The first step appears to be conversion of formaldehyde hydrate to 1-hydroxy-2-oxapropane-3-thiol.

$$H_2S + 2CH_2(OH)_2 \rightleftharpoons HO-CH_2-O-CH_2-SH + H_2O .$$

The second step is visualized as reaction of this product with hydrogen sulfide to form mercaptomethanol, which reacts rapidly

$$HO-CH_2-O-CH_2-SH + H_2S \longrightarrow 2HS-CH_2-OH$$

with oxapropanethiol to form 1-hydroxy-2-oxa-4,6-dithioheptane-7-thiol, the first of these intermediates that is isolated.

$$HO-CH_2-O-CH_2-SH + HO-CH_2-SH \rightleftharpoons$$
$$\rightleftharpoons HO-CH_2-O-CH_2-S-CH_2-SH + H_2O$$
$$HO-CH_2-O-CH_2-S-CH_2-SH + HO-CH_2-SH \rightarrow$$
$$\rightarrow HO-CH_2-O-CH_2-S-CH_2-S-CH_2-SH + H_2O.$$

The dithioheptane continues to react with hydrogen sulfide if a catalytic amount of acid is present. Eventually a product is formed that contains 56.5–60% sulfur and melts at 220–250° C. X-ray diffraction patterns of the product contain spacings indicative of a polythiomethylene and annealing results in weight loss, narrowing of the melting range to 240–255° C, and an increase in sulfur content to over 66%. The theoretical amount of sulfur in poly(thioformaldehyde) is 69.56%. The 66% sulfur product is a highly crystalline, white solid that is insoluble in ordinary organic solvents.

If 1-hydroxy-2-oxa-4,6-dithioheptane-7-thiol is isolated and heated under vacuum, formaldehyde and hydrogen sulfide are eliminated and poly(thioformaldehyde) is formed. This can be done by heating the dithioheptane at 210° C for 6 h *in vacuo*.

$$nHOCH_2OCH_2SCH_2SCH_2SH \longrightarrow 2nCH_2O + (n-1)H_2S + HS(CH_2S)_{2n}H.$$

Typical polymers made in this way melt at 227–237° C and contain 67.7% sulfur *(9)*.

Poly(thioformaldehyde) can also be obtained by reaction of hydrogen sulfide with formaldehyde in acid media *(8)*. Again the key intermediate is 1-hydroxy-2-oxa-4,6-dithioheptane-7-thiol. Though acid conditions favor formation of trithiane, use of a high CH_2O/H_2SO_4 ratio results only in polymer formation. Credali and Russo *(8)* believe this to be a topochemical reaction between the dithioheptane and mercaptomethanol.

$$HOCH_2OCH_2SCH_2SCH_2SH + nHOCH_2SH \longrightarrow HOCH_2O(CH_2S)_{n+3}H + nH_2O.$$

Alkaline solutions of hydrogen sulfide and formaldehyde also give poly(thioformaldehyde). Products containing up to 66.5% sulfur have been obtained from aqueous solutions containing sodium sulfide and formaldehyde in which starting formaldehyde content is 12 weight %. As is shown in Table 1, products with highest sulfur content are obtained when the sodium sulfide is in large excess.

When the formaldehyde concentration at the start is high (36 weight %), the product is a mixture of poly(formaldehyde) and a formaldehyde/thioformaldehyde copolymer.

Table 1

Na$_2$S·9H$_2$O/CH$_2$O at start	pH at start	% S	mp, °C
0.44	11.5	55.5	123—132
1.00	11.5	55.7	125—132
2.93	12.5	66.5	170—180

Polymers with the highest melting points have been obtained by cationic ring-opening polymerization of trithiane (10). Cationic initiators that have been employed include boron trifluoride, methyl iodide, antimony pentafluoride, and dimethyl sulfate. Birkner and Stuerzenhofecker (11) have prepared poly(thioformaldehyde) melting as high as 256° C by the action of the etherate of boron trifluoride on trithiane. They state that polymer made in this way can be molded into a hard plate, that it has a high degree of temperature stability and that it is resistant to acids, alkalies, and oxygen.

A more interesting case is solid state polymerization of trithiane. Stannett (12) has converted trithiane crystals in the solid state to polymer by irradiation with γ-rays from ^{60}Co at 180 or 195° C. Crystals have also been converted to poly(thioformaldehyde) by first irradiating at room temperature and then heating to 180° C (13). Polymer forms during the heating step. The highest melting polymers are obtained from the larger crystals. Poly(thioformaldehyde) made in this way has the same crystal structure (14) as products made from methanedithiol or hydrogen sulfide and formaldehyde.

Nadkarni and Schultz (15) showed that the rate at which polymerization of trithiane is initiated by radiation is a maximum near 175° C and that entropy rate of change for the monomer is also a maximum near 175° C. Further, they established that trithiane undergoes a hitherto unknown phase transition at 194° C.

Two models were proposed to rationalize these interrelationships. One is a molecular motion chain termination model in which it would be presumed that molecular motion would involve orientation involving either coordinated flipping of the ends of the chair conformation trithiane or a rotational motion of the molecule.

Such motion would increase with temperature and would result in molecules becoming placed in a more favorable position for polymerization. However, termination increases with temperature, which would limit polymer molecular weight. The second model involves visualization of the molecules as being in a higher-temperature phase and a lower-temperature one. As temperature increases, the population of the higher-temperature phase increases and this phase is responsible for polymer formation. Neither of these models satisfactorily accommodates all the facts.

Poly(thioformaldehyde) can be made from materials other than formaldehyde and hydrogen sulfide. The reaction of alkaline sulfides or hydrosulfides with methylene chloride is an interesting case. Schmidt and Blaettner (16) were the first to examine this reaction, but they did not characterize the product. Later, Russo et al. (17), found that poly(thioformaldehyde) can be made from sodium hydrosulfide and methylene chloride in yields of over 90%

$$NaSH + CH_2Cl_2 \longrightarrow +CH_2S\dashv_n + NaCl + H_2S.$$

Products obtained in this way had melting points from 160 to 205° C. These relatively low melting points indicate reduced molecular weight, which was also attested to by the presence of a fraction soluble in hot benzene. The benzene-insoluble part melted around 230° C.

This polymer has a crystal structure different from the normal hexagonal poly(thioformaldehyde). At 153° C it changes to the hexagonal configuration. Credali and Russo (8) state that this new form can be heat-stabilized by acetylation of its end groups and that the acetylated polymer does not degrade at 250–260° C, a temperature range at which ordinary poly(thioformaldehyde) is very unstable.

In addition to the polymer, the reaction of alkaline sulfides and hydrosulfides with methylene chloride gives trithiane (16), 1,3,5,7-tetrathiocane (16), and 1,3,5,7,9-pentathiacyclodecane (18).

Reaction of bis(chloromethyl)sulfide with sodium sulfide (19) gives poly(thioformaldehyde) of melting point 220–245° C. If the reaction is carried out in water-alcohol solution, the cyclic compounds, 1,3,5,7,9-oxatetrathiacyclodecane (I) and 1,3,5,7,9,11-oxapentathiacyclododecane (II) (20), are formed.

From the discussion it is obvious that melting temperature of poly(thioformaldehyde) depends upon method of preparation. Those melting points range

Table 2. Melting Points of $HS(CH_2S)_nH$

n	mp, °C
2	70— 80
4	110—120
14	200—210
20	210—220
25—30	240

from 220–260° C (5, 8, 10). This variation is due in part to copolymerization with formaldehyde in those cases in which polymer is made from formaldehyde and hydrogen sulfide. Melting point also is related to the number of thiomethylene groups in the chain (21) as shown in Table 2.

C. Toxicity

A word of caution about toxicity. Many of these compounds are so malodorous it is unlikely anyone could tolerate lethal concentrations of them. Just the same, one should remember that they are toxic and that exposure to them should be minimized. They should only be handled in hoods with good drafts. On human skin, poly(thioformaldehyde) causes dermatitis. Its LD_{50} in rats is 365 ± 11 mg/kg (21).

D. Crystal Structure

As mentioned earlier, poly(thioformaldehyde) is a highly crystalline polymer, and it has been shown to have a hexagonal crystal structure (14, 24). Thus, poly-(thioformaldehyde) obtained by irradiation of trithiane followed by heating has a hexagonal unit cell with $a = 5.07$ Å and $c = 36.52$ Å crossed by one helical chain parallel to the c axis. The chain has an identity period of 36.52 Å that comprises 17 —CH_2S— units in 9 turns of the helix (14). Poly(thioformaldehyde) made by other methods crystallizes similarly (24).

Inasmuch as trithiane forms orthorhombic crystals, it was first thought that the polymer from it also was orthorhombic (13). Indeed, it may be described by means of a pseudo-orthorhombic unit cell (14) in which $a = 5.07$ Å, $b' = 8.77$ Å, and $c = 36.52$ Å. Three —CH_2S— groups in the helix along the c axis have a spacing of 6.45 Å, i.e., $c' = 6.45$ Å. By comparison with a trithiane unit cell, which is orthorhombic with $a = 7.668$ Å, $b = 7.003$ Å, $c = 5.285$ Å, and six —CH_2S— groups, and by making a, b', and c' of the polymer correspond with c, a, and b of trithiane, respectively, it can be concluded that solid state polymerization of trithiane could occur along its b axis (14). In fact, Lando and Stannett (13) state, "... polymerization of crystalline trithiane is seen to be a good example of a topotactic process". However, they feel that most of the polymer grows along the ab diagonals of trithiane.

A thorough study of the crystal structure of hexagonal poly(thioformaldehyde) has been made by Carazzolo and Valle (24). They have compared poly(thioformaldehyde) to hexagonal poly(formaldehyde) (25). This is one of two forms of poly(formaldehyde), the other being orthorhombic. These authors give the data for poly(thioformaldehyde) listed above, which is 17 —CH_2S— units and 9 turns of the helix in the unit cell. They also state that the radius of the helix is 0.99 Å, as compared to 0.68 Å in hexagonal poly(formaldehyde), and that the internal rotation angle is 65° 59'. Thus, the CH_2 and S are more staggered as one looks down on the helix than is the case for hexagonal poly(formaldehyde), in which case the internal rotation angle is 76° 14'.

E. Applications

Though much effort has been spent on possible utilization of poly(thioformaldehyde), no uses for the polymer have been developed. Most samples of the polymer cannot be melt fabricated because they are unstable in the molten state. Solvent fabrication is also impossible because the polymer is insoluble. There are two types of poly(thioformaldehyde) that are said to be stable at high temperatures, which are polymers made by treatment of trithiane with boron trifluoride etherate and the acetylated anomalous form obtained from sodium hydrosulfide and methylene chloride. These also have not found utility.

A number of modified poly(thioformaldehyde) compositions have been examined. These include a "liquid" poly(thioformaldehyde) made from hydrogen sulfide and aqueous formaldehyde at 80° C (26). Its composition is unknown but it contains up to 73% sulfur. It is said to be a rubber vulcanizing agent. Low molecular weight polymers of degree of polymerization 3–7 have been condensed with ketones, phenol, or urea to give products suitable for floor and wall coatings (27). Poly(thioformaldehyde) has also been heated with large amounts of sulfur to give a plastic of good abrasion resistance (28). Low molecular weight polymers of thioformaldehyde have also been added to 1,3-dienes to give compositions in which each end of the chain is capped (29).

$$HS\!-\!(CH_2S)_{2-8}\!H + CH_2\!=\!CH\!-\!CH\!=\!CH_2 \longrightarrow$$
$$CH_2\!=\!CH\!-\!CH_2CH_2S\!-\!(CH_2S)_{2-8}SCH_2CH_2CH\!=\!CH_2.$$

III. Thioacetone

A. Introduction

After thioformaldehyde, the thiocarbonyl compound that has received most attention is thioacetone. Unlike thioformaldehyde, it can be kept in monomeric form, but only if kept below −50° C (30). It is a red oil that freezes at about −55° C and boils at about 70° C. If pure it polymerizes to a white solid in a few hours at −78° C or quite rapidly at room temperature. It also exists as a thioenol tautomer, which is stable at temperatures below −50° C but which tautomerizes at higher temperatures to give an equilibrium mixture of thioketo and thioenol forms.

B. Preparation

Thioacetone was first made in 1889 by Baumann and Fromm (*31*). They obtained it by reaction of hydrogen sulfide with acetone in the presence of an acidic catalyst. It was a minor product detected by its very disagreeable odor. The major product was hexamethyl-s-trithiane. In more recent work (*32, 33*) pyrolysis of this intermediate has been developed into a preferred route to thioacetone.

Reaction of acetone with hydrogen sulfide in the presence of acidified $ZnCl_2$ at 25° C gives a good yield of a product composed of 60–70% hexamethyltrithiane and 30–40% of 2,2-propanedithiol. Thioacetone can be obtained by pyrolysis of either of these compounds. The trithiane is pyrolyzed either on quartz rings heated to 500–650° C at 5–20 mm (*30*) or by means of a hot wire (*32*). The dithiol is pyrolyzed on sodium fluoride pellets heated to 150° C at 11 mm (*30*). In both cases the pyrolysate is immediately collected in a trap cooled to $-78°$ C.

Pyrolysis of allyl isopropyl sulfide is reported

$$CH_2\!=\!CHCH_2\!-\!S\!-\!\underset{\underset{CH_3}{|}}{\overset{\overset{CH_3}{|}}{CH}} \xrightarrow{\Delta} CH_2\!=\!CHCH_3 + \underset{CH_3}{\overset{CH_3}{\diagdown}}\!\!C\!=\!S$$

by Bailey and Chu (*34*) to give a product that is solely thioacetone. Thioacetone has also been prepared by cleavage of an acetone ketal with hydrogen sulfide (*35, 36*). Material made in this way was not obtained in pure form.

The reaction of acetone and hydrogen sulfide gives, in addition to hexamethyl-s-trithiane, small amounts of two isomeric impurities, 2,3,5,5,6,6-hexamethyl-1,2,4-trithiane and 4-mercapto-2,2,4,6,6-pentamethyl-1,3-dithiane (*32*).

The mercapto impurity can be removed by stirring the mixture with benzoquinone before pyrolysis.

C. Thioketo and Thioenol Tautomers

Careful examination of the pyrolysis of the trithiane from thioacetone has shown that the product is always a mixture of the thioketone, CH_3CSCH_3, and the thioenol, $CH_2\!=\!CSHCH_3$ (30). The amount of thioenol, which is always the lesser component, varies from about 5–15% depending upon pyrolysis conditions. Below 500° C, the trithiane is completely pyrolyzed. Above 600° C, thioacetone undergoes extensive decomposition. In between 500° C and 600° C, the relative proportion of thioketo tautomer increases with increasing pyrolysis temperature.

The thioketo tautomer in the pyrolysis products is detectable by infrared. It shows strong absorption at 7.85 μ, attributable to the C=S group. This absorption is at shorter wavelengths than that assigned to C=S absorption in more complex thioketones (37, 38). The thioenol tautomer absorbs at 3.93 μ for SH, at 6.15 μ for C=C, and at 11.55 μ for =CH_2.

Spectra of these tautomers were obtained on samples separated by gas chromatography from thioketo-thioenol mixtures. Trithioacetone pyrolysate was passed through a DC 200 silicone oil on Gas Chrom Z packed column. As the appropriate sample eluted from the column, it was condensed on a sodium chloride plate cooled to −100° C and its spectrum immediately determined. The very low temperature employed prevented loss of thioenol by tautomerization and of thioketo by polymerization.

NMR characterization of the tautomers has shown that the pure keto isomer just above its melting point gives a single peak. This, of course, is required if structure assignment is correct. Absorption is at −2.67 ppm with reference to tetramethylsilane as an internal standard. This tautomer is extremely reactive and polymerized to an appreciable extent even at −78° C in a few hours or in a few seconds at room temperature.

The NMR spectrum of the pure enol in deuterochloroform solution at −52° C contains three proton resonances. One is at −4.93 ppm for vinylic protons, a second at −2.8 ppm for SH, and a third at −1.95 ppm for methyl protons, all with reference to tetramethylsilane as an internal standard. The pure enol tautomer is much less reactive than the thioketo form as it does not change appreciably after one hour at room temperature. It does undergo deep seated changes, however, when kept a few days at room temperature.

As mentioned earlier, pyrolysis of 2,2-propanedithiol over sodium fluoride at about 200° C also leads to thioacetone. The product obtained at 150° C is more than 90% the thioketo form. At 250° C, the thioenol amounts to over 50% of the product.

D. Polymerization

As has been mentioned earlier, thioacetone polymerizes adventitiously even at very low temperatures. The product is a white powder of very high crystallinity. Bailey and Chu (34) state that pure thioacetone gives only polymer, but others always have obtained hexamethyl-s-trithiane in addition to polythioacetone. The difference may be in method of preparation of the monomer. Thioacetone employed by Bailey and Chu was obtained solely by pyrolysis of allyl isopropyl sulfide. Other investigators have used mainly pyrolysis of hexamethyl trithiane to generate thioacetone. Polymer from monomer prepared either way had the same infrared and NMR absorption capabilities. Infrared absorption occurs at 2950, 2900, 1440, 1150, 1360, and 1375 cm^{-1} for the *gem*-dimethyl group and at 1085 and 643 cm^{-1} for C—S. The NMR spectra of all samples contain a single peak at $\tau = 8.1$.

Melting point of poly(thioacetone) varies considerably depending upon method of polymerization and source of the monomer. Polymer melting point has been related to molecular weight. Bailey and Chu give a melting point range of 71–116° C. The state that poly(thioacetone) melting at 101–105° C has a molecular weight of 2000. Ettingshausen and Kendrick (32) report a melting point range of 120–125° C. They also have related spontaneous polymerization temperature and molecular weight as follows:

Polymerization temperature, °C	Molecular weight
20	2500
0	6500
−10	9200
−20	14000

Using thioacetone prepared by pyrolysis of hexamethyl-s-trithiane in a modified ketene lamp, Burnop and Latham (33) made a study of spontaneous polymerization of this monomer. By collecting the monomer at −78° C and allowing it to warm up undisturbed, they obtained a 72% yield of poly(thioacetone) melting at 119–120° C. In an equal volume of ether, thioacetone polymerized to give an 81% yield of polymer melting at 124–125° C. The solution was prepared at −78° C and allowed to warm slowly. At −30° C to −25° C cloudiness appeared indicating polymerization. In ethylene oxide, yields as high as 90% of polymer melting at 122–123° C were obtained.

The studies described above did not take into account thioenol content. Polymer formed in its presence by either an ionic mechanism or a free-radical mechanism would probably be of lower molecular weight. It was found (30) that thioacetone of greater than 95% thioketo tautomer gives polymer melting at 120–124° C, which is in line with the other products.

Free-radical initiation at low temperature has been shown to promote polymerization of thioacetone (30). Initiation with the triethylborane-diethyl(peroxyethyl)borane redox couple (39), the behavior of which is discussed later, at 50° C gave polymer of high molecular weight as judged by melting point. It did not, however, have the toughness expected of a high molecular weight polymer. A somewhat lower molecular weight product, also weak and brittle, was obtained by ultraviolet irradiation of the thioketo tautomer. Both polymers were stable for several weeks in contrast to polymer prepared from thioacetone containing considerable thioenol.

Polymerization of thioacetone is also initiated by visible light. In fact, it has been shown that exposure of frozen crystals of thioacetone monomer to light leads to rapid solid phase polymerization (33).

Attention has been given to initiation of the polymerization of thioacetone by bases. When small amounts of potassium *tert*-butoxide are added to thioacetone, its orange color is rapidly lost and solid polymer is slowly formed. The polymer must be purified or else it gradually changes to a viscous liquid. Polymer formed in this way is low melting (71–73° C) and is unstable as it slowly depolymerizes. It has been postulated that base favors formation of the thioenol which leads to low molecular weight polymer and is responsible for introducing thiol and vinyl groups in the polymer.

Though thioacetone polymerizes spontaneously, by free-radical initiation, and by base initiation, it does not form copolymers in any of these systems. Efforts to obtain copolymers with dienes, vinyl compounds, acrylates, aldehydes, and epoxides have failed.

IV. Higher Thiocarbonyl Compounds

A. Introduction

A number of higher thioketones have been prepared by pyrolysis of hexasubstituted trithianes (40, 41). These trithianes were made by methods already discussed for synthesis of hexamethyl-*s*-trithiane. Those investigated include compounds of the structure

where R and R' are alkyl, aryl, aralkyl, cycloalkyl, or heterocyclic (40). Specific thioketones reported are methyl ethyl thioketone, diethyl thioketone, diisopropyl thioketone, methyl benzyl thioketone, ethyl amyl thioketone, ethyl isoamyl thioketone, and thioacetophenone. Of these, the only polymer discussed in any detail is that from thioacetophenone (41).

B. Thioacetophenone

Thioacetophenone is readily obtained by pyrolysis of 2,4,6-trimethyl-2,4,6-triphenyl-1,3,5-trithiane at 190–195° C (41). It is a deep purple liquid (λ_{max} 310 mu, $\varepsilon = 10900$) that solidifies at about $-25°$ C. Because it is difficult to handle, it has been used directly as prepared without additional purification. The infrared spectrum of thioacetophenone shows strong absorption at 1260 cm^{-1} for C=S. The NMR spectrum is described as consisting of a sharp methyl singlet at 3.05 ppm and two groups of aromatic multiplets at 7.90–8.20 ppm and 7.00–7.55 ppm. No evidence for a thioenol tautomer has been obtained in this case.

Thioacetophenone polymerizes spontaneously to a white solid. Conversion to polymer is increased by irradiation with ultraviolet light. Highest conversion reported is 74%. Polymerization is also brought about by such anionic initiators as $AlEt_2Cl$, $AlEt_3$, $BF_3 \cdot OEt$, and $SnCl_4$. Molecular weights measured on tetrahydrofuran or benzene solutions using a vapor-pressure osmometer were in the 1000–2000 range. These are minimum values, since the polymer decomposes readily.

Upon standing for a few weeks, poly(thioacetophenone) gradually decomposes to a purple, viscous liquid. Decomposition is hastened by heating.

C. 1-Thioacylaziridines

The polymerization of 1-thioacylaziridines represents a special case as these compounds polymerize 1,5 exclusively (42).

$$C_6H_5-\underset{\underset{S}{\|}}{C}-N\underset{CH_2}{\overset{CH_2}{\diagup\!\!\!\diagdown}} \longrightarrow \{CH_2CH_2-N=\underset{\underset{C_6H_5}{|}}{C}-S\}_n.$$

1-(Thiobenzoyl)aziridine polymerizes very rapidly at room temperature. 1-(p-Methylthiobenzoyl)aziridine and 1-(p-methoxythiobenzoyl)aziridine polymerize much more slowly.

Solution polymerization of these compounds can be brought about by nucleophilic initiators including n-butyllithium, triethylamine, and sodium cyanide. In the absence of such initiators, solution polymerization proceeds very slowly. As an example, 1-(p-chlorothiobenzoyl)aziridine at a concentration of 0.5 mole percent in tetrahydrofuran polymerizes at room temperature when initiated with n-butyllithium to give a 94% yield of polymer. Melting point of the polymer is 90–100° C and its reduced viscosity in N-methylpyrrolidone (1% concentration at 30° C) is 0.15.

1-Thiobenzoylaziridine polymerizes similarly and the polymer obtained melts in about the same neighborhood (84–98° C). Somewhat higher melting products are formed by 1-(p-methylthiobenzoyl)aziridine (142–153° C) and 1-(p-methoxythiobenzoyl)aziridine (125–127° C).

1-(N-Alkyl or arylthiocarbamyl)aziridines also polymerize 1,5 (41). Reduced viscosities of the polymers (1% in N-methylpyrrolidone at 30° C) of up to 0.2–0.3

were observed for the polymer

$$C_6H_5-NH-\underset{\underset{S}{\|}}{C}-N\underset{CH_2}{\overset{CH_2}{<}} \longrightarrow (CH_2CH_2-N=\underset{\underset{NHC_6H_5}{|}}{C}-S)_n$$

$$\text{or} \quad (CH_2CH_2-NH-\underset{\underset{NC_6H_5}{\|}}{C}-S)_n$$

from 1-(N-phenylthiocarbamyl)aziridine. It is a poly(isothiourea) as indicated by a C=N stretching band in the infrared at 1580 cm^{-1} and by contrast with authentic poly-[N-(N'-phenylthiocarbamyl)ethyleneimine], which is the product that would result from 1,3-polymerization and which was prepared for comparison from polyethyleneimine and phenylisothiocyanate.

Poly(iminothiocarbonates) and poly(iminodithiocarbonates), both of which are terminated with an isocyanate group, are obtained from 1-(aryloxythiocarbonyl)aziridines and 1-(aryldithiocarbonyl)aziridines, respectively.

$$ArOC-N\underset{CH_2}{\overset{CH_2}{<}} \longrightarrow (CH_2CH_2-\underset{\underset{OAr}{|}}{C}=N-S)_nCH_2CH_2N=C=S$$

$$ArSC-N\underset{CH_2}{\overset{CH_2}{<}} \longrightarrow (CH_2CH_2\underset{\underset{SAr}{|}}{C}=N-S)_nCH_2CH_2N=C=S$$

Post reaction of these polymers with aniline leads to polymers end capped with a thiourea group. Aziridines of these types that have been polymerized include p-tolyl and p-nitrophenyloxythiocarbonyls and p-tolyl, p-chlorophenyl and p-methoxyphenyldithiocarbonyls. Molecular weights in all cases were low, being in the neighborhood of 1500–3000.

V. Fluorothiocarbonyl Compounds

A. Introduction

Fluorine-substituted thiocarbonyl compounds have been studied even more intensively than thioformaldehyde and thioacetone. These compounds have a very rich chemistry of which polymerization is only a part. The simplest member of this class is fluorothiocarbonyl fluoride, $CF_2=S$, which also forms the most interesting polymers. Other members that have been investigated include a variety of fluorothioacyl halides and a number of fluorothioketones. Because preparative methods for obtaining these compounds embrace a versatile segment of chemistry, synthesis of monomers is discussed first and in some detail. Polymers have been obtained from these compounds by ionic initiation. In the case of thiocarbonyl fluoride and thiocarbonyl chlorofluoride, free-radical initiation is also effective. Polymer formation is discussed under each of these headings, with ionic methods coming first because they apply to the widest range of fluorothiocarbonyl compounds.

B. Synthesis of Monomers

1. Thiocarbonyl Fluoride

As indicated above, the most important of the thiocarbonyl compounds is thiocarbonyl fluoride, $CF_2=S$. It has been synthesized by a number of routes (43), the most convenient of which for laboratory investigation is dimerization of thiophosgene followed by conversion of the dimer to 2,2,4,4-tetrafluoro-1,3-dithietane (I) and thermal cracking of the dithietane (44).

$$CCl_2=S \underset{}{\overset{h\nu}{\rightleftharpoons}} Cl_2C\underset{S}{\overset{S}{\diamond}}CCl_2 \longrightarrow F_2C\underset{S}{\overset{S}{\diamond}}CF_2 \overset{500°\,C}{\longrightarrow} CF_2=S$$

I

A detailed procedure for preparing thiocarbonyl fluoride by this route has been published by Sharkey and Jacobson (45).

The replacement of chlorine in the dithietane by fluorine occurs stepwise. As a consequence, small amounts of the partially fluorinated dithietanes II, III, and IV are ordinarily obtained. Since efforts are usually made to fluorinate

$$Cl_2C\underset{S}{\overset{S}{\diamond}}CClF \qquad FClC\underset{S}{\overset{S}{\diamond}}CFCl \qquad F_2C\underset{S}{\overset{S}{\diamond}}CFCl$$

II   III   IV

as completely as possible, in most cases the only one of these obtained in any significant quantity is IV. Thermal cracking of this dithietane leads to a mixture of thiocarbonyl chlorofluoride and thiocarbonyl fluoride (39). These two can be separated by fractionation. Thiocarbonyl fluoride is a colorless gas boiling

$$F_2C\underset{S}{\overset{S}{\diamond}}CFCl \overset{600°\,C}{\longrightarrow} CF_2=S + CFCl=S$$

at $-54°$ C, and the chlorofluoride is a bright yellow liquid boiling at 8–10° C.

The configuration of the thiocarbonyl fluoride molecule has been determined by spectral studies. In the infrared (46, 47) absorption occurs at 1368 cm^{-1} for C=S, 787 cm^{-1} for symmetrical C—F stretching, 526 cm^{-1} for symmetrical CF$_2$ bending, 1189 cm^{-1} for asymmetrical C—F stretching, 417 cm^{-1} for asymmetrical CF$_2$ bending, and 622 cm^{-1} for out-of-plane bending. The photoelectron spectrum (48) of thiocarbonyl fluoride consists of a number of strong bands, four of which exhibit well resolved vibrational structure. The first of these, which has an adiabatic ionization potential of 10.45±0.01 eV, has been assigned to ionization of the sulfur 3p[b$_2$] electron. The second band has an adiabatic ionization potential of 11.34±0.01 eV and has been assigned to ionization of an electron from the C=S Π [b$_1$] bonding orbital. The C—F distance, C=S bond length, and F—C—F bond angle have been determined from microwave spectra (49) to be 1.315±0.01 Å, 1.589±0.01 Å, and 107.1°±1.0, respectively. In this same study, dipole moment was determined to be 0.080 D.

A number of routes to thiocarbonyl fluoride that do not involve tetrafluorodithietane have been developed. In one (50), phosgene is chlorinated to give trichlorosulfenyl chloride, which is converted to chlorodifluorosulfenyl chloride by reaction with antimony trifluoride, and the fluorinated compound is then dehalogenated by reaction with tin.

$$CCl_2{=}S \xrightarrow{Cl_2} CCl_3SCl \xrightarrow{SbF_3} ClCF_2SCl \xrightarrow{Sn} CF_2{=}S.$$

Thiocarbonyl fluoride has also been prepared by reaction of bis(trifluoromethylthio)mercury and iodosilane (51). The first product formed is trifluoromethylthiosilane, which spontaneously decomposes to thiocarbonyl fluoride and fluorosilane.

$$(CF_3S)_2Hg + SiH_3I \longrightarrow SiH_3(SCF_3) + HgI_2.$$
$$\downarrow$$
$$CF_2{=}S + FSiH_3$$

A direct route to thiocarbonyl fluoride is reaction of chlorodifluoromethane with sulfur in a hot tube at 700–900° C (52). This reaction gives very high yields and is amenable to preparation of

$$ClCF_2H + S \xrightarrow{700-900°\,C} CF_2{=}S + HCl$$

large amounts of monomer. Thiocarbonyl fluoride prepared in this way is contaminated with large amounts of HCl. The HCl can be separated by fractionation since thiocarbonyl fluoride boils considerably higher than anhydrous HCl.

Another direct route is reaction of tetrafluoroethylene with gaseous sulfur in a hot tube (53). The main product is thiocarbonyl fluoride. Lesser amounts of trifluorothioacetyl fluoride and bis(trifluoromethyl)disulfide are also formed.

$$CF_2{=}CF_2 + S \xrightarrow{500-600°\,C} CF_2{=}S + CF_3CF{=}S + CF_3SSCF_3.$$

2. Fluorothioacyl Halides

The reaction of a fluoroolefin with sulfur is a somewhat general preparative method for making fluorothioacyl halides. When the reaction is conducted over a bed of activated charcoal (53), trifluorothioacetyl fluoride is formed from tetrafluoroethylene in high yields. Charcoal is not needed in the case of chloro- and bromofluoroethylenes, which react with sulfur at high temperatures to give high yields of chlorodifluorothioacetyl fluoride and bromodifluorothioacetyl fluoride, respectively. It has been suggested (44) this transformation involves formation of an episulfide that rearranges by migration of a chlorine or bromine atom. The fact that 1,1-dichloro-2,2-difluoroethylene gives chlorodifluorothioacetyl chloride supports this hypothesis.

$$CF_2{=}CCl_2 \xrightarrow[445°C]{S} \left[\underset{Cl}{\overset{S}{CF_2{-}C{-}Cl}} \right] \longrightarrow ClCF_2\overset{S}{\overset{\|}{C}}Cl.$$

Chlorodifluorothioacetyl fluoride prepared by this reaction has a boiling point of 23° C. NMR characterization of this compound gave a low-field triplet and high-field doublet in a ratio of 1:2, as would be expected. However, a material boiling at 36° C had been reported earlier as chlorodifluorothioacetyl fluoride (54). This material was obtained by dehalogenation of 2-chlorotetrafluoroethanesulfenyl chloride with tin. In view of

$$ClCF_2CF_2SCl \xrightarrow{Sn} ClCF_2\overset{S}{\overset{\|}{C}}F$$

the latter results, it seems likely this product was contaminated with higher-boiling, difficultly-removable substances that contributed to an artificially high boiling point.

Sulfur also reacts with fluoroalkyl mercurials and fluoroalkyl iodides to give fluorothioacyl halides. In the case of the reaction with mercurials, the halide formed is determined by substitution on the carbon attached to mercury. For example, bis(perfluoroethyl)mercury gives trifluorothioacetyl fluoride and bis-(1,1-dichloro-2,2,2-trifluoroethyl)mercury gives trifluorothioacetyl chloride.

$$(CF_3CF_2)_2Hg \xrightarrow[445°C]{S} CF_3\overset{S}{\overset{\|}{C}}F$$

$$(CF_3CCl_2)_2Hg \xrightarrow[445°C]{S} CF_3\overset{S}{\overset{\|}{C}}Cl$$

Reaction of elemental sulfur with fluoroalkyl iodides has not received very much attention, but it has been shown that sulfur converts 4,4-diiodoperfluoro-1-butene to pentafluoro-3-butenoyl fluoride.

$$CF_2=CFCF_2CFI_2 \xrightarrow[450°C]{S} CF_2=CFCF_2\overset{\overset{\displaystyle S}{\|}}{C}F$$

A very versatile route to fluorothiocarbonyl compounds is dehydrohalogenation of fluorine-containing mercaptans. In the case of removal of HF, sodium fluoride is a very convenient dehydrohalogenating agent.

$$CF_3CF_2SH + NaF \longrightarrow CF_3\overset{\overset{\displaystyle S}{\|}}{C}F + NaHF_2$$

Harris and Stacey (55) have combined this with free-radical addition of hydrogen sulfide to fluoroethylenes as a method for preparing fluorothioacetyl fluorides.

$$ClCF=CF_2 + H_2S \xrightarrow{X\text{-ray}} ClCFHCF_2SH$$

$$ClCFHCF_2SH + NaF \longrightarrow ClCFH\overset{\overset{\displaystyle S}{\|}}{C}F$$

In addition to chlorofluorothioacetyl fluoride, difluorothioacetyl fluoride and fluoromethoxythioacetyl fluoride have been made by

$$HCF_2\overset{\overset{\displaystyle S}{\|}}{C}F \qquad CH_3OCF\overset{\overset{\displaystyle S}{\|}}{C}F$$

this route.

3. Fluorothioketones

With slight modification, the methods used to prepare fluorothioacyl fluorides can also be used for synthesis of fluorothioketones. Hexafluorothioacetone, the member of this class that has been studied most extensively, is readily obtained by high-temperature reaction of hexafluoropropylene and sulfur (53). The thioketone is a deep-blue liquid, bp 8° C, that dimerizes on standing to 2,2,4,4-tetrakis-(trifluoromethyl)-1,3-dithietane.

$$CF_3CF=CF_2 \xrightarrow[\Delta]{S} \underset{\underset{\displaystyle CF_3}{|}}{\overset{\overset{\displaystyle CF_3}{|}}{C}}=S \longrightarrow \begin{array}{c} CF_3 \\ | \\ C \\ | \\ CF_3 \end{array} \overset{S}{\underset{S}{\diagdown\diagup}} \begin{array}{c} CF_3 \\ | \\ C \\ | \\ CF_3 \end{array}$$

It is best to store the thioketone as the dithietane, which is a white solid melting at 23.6° C and boiling at 110° C. When needed, hexafluorothioacetone can be obtained from the dithietane by pyrolysis in a hot tube at 600° C (44). Pyrolysis gives a quantitative yield of monomer. The thioketone so prepared is purified by distillation at 200 mm at which pressure it boils at $-20°$ C.

The reaction of a perfluoromercurial with sulfur is particularly well suited to synthesis of hexafluorothioacetone. The first step is addition of mercuric fluoride to hexafluoropropylene in liquid HF which gives bis(hexafluoroisopropyl)mercury.

$$CF_3CF{=}CF_2 + HgF_2 \xrightarrow{HF} (CF_3)_2CF{-}Hg{-}CF(CF_3)_2$$

Passage of the mercurial through the vapor of boiling sulfur results in its conversion to hexafluorothioacetone in 60% yield (56).

$$(CF_3)_2CF{-}Hg{-}CF(CF_3)_2 + S \xrightarrow{445°C} CF_3{-}\underset{\|}{\overset{S}{C}}{-}CF_3 + HgF_2$$

At lower temperatures, the reaction results in formation of a mixture of hexafluoroisopropyl di- and polysulfides.

$$(CF_3)_2CF{-}Hg{-}CF(CF_3)_2 \xrightarrow[200°C]{S} (CF_3)_2CF{-}(S)_n{-}CF(CF_3)_2$$

$$(n = 2, 3, \text{ or } 4)$$

The disulfide can be defluorinated by triphenylphosphine.

The hexafluorothioacetone is isolated as its dimer, 2,2,4,4-tetrakis(trifluoromethyl)-1,3-dithietane, which forms immediately. In the case of the tri- and tetrasulfides, excess triphenylphosphine is used to remove the extra sulfur atoms by formation of triphenylphosphine sulfide.

A recent study (57) indicates the disulfide can also be formed by reaction of hexafluoropropylene with potassium fluoride and sulfur. It appears that the key intermediate in this reaction is the anion $(CF_3)_2CFS^\ominus$. 2,2,4,4-Tetrakis(trifluoromethyl)-1,3-dithietane and $(CF_3)_2CFSCF=C(CF_3)(CF_2)_2CF_3$ are also said to be formed.

Secondary perfluoroalkyl iodides react with sulfur to form disulfides, which can be reduced by ultraviolet irradiation in the presence of a higher boiling thiol. Treatment of the

$$C_2F_5CFICF_3 \xrightarrow{S} \underset{\underset{CF_3}{|}}{\overset{\overset{C_2F_5}{|}}{FC}}\!\!-\!\!S\!-\!\!S\!-\!\!\underset{\underset{CF_3}{|}}{\overset{\overset{C_2F_5}{|}}{CF}} \underset{\rightleftharpoons}{\overset{RSH}{}} \underset{\underset{CF_3}{|}}{\overset{\overset{C_2F_5}{|}}{FC}}\!\!-\!\!SH \xrightarrow{NaF} \underset{\underset{CF_3}{|}}{\overset{\overset{C_2F_5}{|}}{C}}\!\!=\!\!S$$

perfluoromercaptan so formed with sodium fluoride leads to a perfluorothioketone, octafluoromethylethylthioketone in the example given. The iodide is directly convertible to a thioketone by reaction with phosphorus pentasulfide at its reflux temperature. It has been proposed (44) that this transformation involves the following steps:

$$\underset{\underset{CF_3}{|}}{\overset{\overset{C_2F_5}{|}}{FCI}} \longrightarrow \underset{\underset{CF_3}{|}}{\overset{\overset{C_2F_5}{|}}{FC\cdot}}$$

$$P_2S_5 \rightleftharpoons P_2S_3 + S_2$$

$$\underset{\underset{CF_3}{|}}{\overset{\overset{C_2F_5}{|}}{FC\cdot}} + S_2 \longrightarrow \underset{\underset{CF_3}{|}}{\overset{\overset{C_2F_5}{|}}{FC}}\!\!-\!\!S\!-\!\!S\!-\!\!\underset{\underset{CF_3}{|}}{\overset{\overset{C_2F_5}{|}}{CF}} \xrightarrow{P_2S_3} \underset{\underset{CF_3}{|}}{\overset{\overset{C_2F_5}{|}}{C}}\!\!=\!\!S$$

C. Fluorothiocarbonyl Polymers by Ionic Initiation

1. Poly(thiocarbonyl Fluoride)

Very high molecular weight polymers are formed by anionic polymerization of thiocarbonyl fluoride at low temperatures (58). These products are thioacetals that come about through addition polymerization of the C=S bond.

$$nCF_2=S \longrightarrow +CF_2-S+_m$$

Polymerization takes place very rapidly at $-78°$ C under anhydrous conditions in aprotic solvents. A detailed procedure for forming polymer in anhydrous ether using dimethylformamide (DMF) as the initiator has been described by Sharkey and Jacobson (45).

The degree of polymerization that is attained is influenced by polymerization temperature, polymerization medium and nature of the initiator (59, 60). Using high purity monomer, anhydrous ether as a solvent, DMF as the initiator, and a polymerization temperature of $-78°$ C, poly(thiocarbonyl fluoride) is formed that has an inherent viscosity (0.5% solutions in chloroform at $250°$ C) in the range of 4–6. Polymers prepared at $-50°$ C under otherwise similar conditions have inherent viscosities around 2. When tetraisopropyl titanate is used as the initiator, the polymers obtained have inherent viscosities of 2–4.5. Intermediate molecular weight products are formed by use of tetraisopropyl titanate in conjunction with isopropyl alcohol as a chain transfer agent. Low molecular weight oils result when relatively large amounts of tetraisopropyl titanate are used and the polymerization is carried out at $-25°$ C in chloroform.

A variety of other anionic initiators promote the polymerization of thiocarbonyl fluoride. Among them are aluminum isopropoxide, di(hydrogenated tallow)dimethylammonium methoxide and chloride, N-nitrosodimethylamine, diisopropylamine, and triethylamine (59). Combinations of either $[(C_6H_5)_3P]_3RhCl$ or

$$\begin{array}{c}(C_6H_5)_3P \diagdown \quad Cl \diagdown \quad P(C_6H_5)_3 \\ Rh \quad Rh \\ (C_6H_5)_3P \diagup \quad Cl \diagup \quad P(C_6H_5)_3\end{array}$$

with AlR_2X (R = butyl, X = butyl, H, or Cl) at an Al–Rh ratio of 1:1–1:10 are also effective catalysts (61). Polymerization is also brought about by ^{60}Co α-irradiation (62). The nature of the initiation in this case is obscure, though it seems likely it is ionic.

Since an inherent viscosity of 2.03 for a 1% solution of the polymer in chloroform corresponds to an $\bar{M}_n$ of 300000 to 400000, it is apparent the upper limit of molecular weight is very high. The high molecular weight polymer is surprisingly resistant to acids and aqueous base. Boiling in fuming nitric acid causes no apparent damage. If boiling is continued for a long time, degradation does occur. For example, boiling for 21 h caused the inherent viscosity of one sample to drop from 3.45 to 0.54. Similarly, boiling in 10% sodium hydroxide solution causes no damage unless continued for a long time. After 21 h, about 9% loss in weight occurred, but inherent viscosity did not change. The polymer is rapidly attacked by amines, however.

Above $175°$ C, poly(thiocarbonyl fluoride) decomposes (63) to regenerate thiocarbonyl fluoride monomer. Chains capped with CF_3 are stable to $300°$ C when heated in nitrogen, though they decompose at $200°$ C or below when heated in air. Chains so capped have been made by reaction of the polymer with antimony pentafluoride.

$$+\!\!(CF_2\!-\!S)_{\overline{m}} \xrightarrow{SbF_5} CF_3S+\!\!(CF_2\!-\!S)_{\overline{\leq 25}}CF_3$$

Degradation also occurs with the result that relatively low molecular weight oils are formed.

Study of infrared and NMR spectra have established that the end groups in poly(thiocarbonyl fluoride) are CF_3S-, $-SCF=S$, and $-SCF=O$. All three end groups are present in polymer initiated by dimethylformamide, though $-SCF=O$ appears to be present in smaller amounts than the other two. Identification of these groups is as follows (64):

Dimethylformamide initiated polymer

Group	NMR	IR
CF_3S-	37.6 ppm higher field	13.1 μ
$-(CF_2-S)_n-$	43.5 ppm higher field	—
$-CF=S$	71.5 ppm lower field	8.2 μ

Tetraisopropyl titanate initiated polymer

Group	NMR	IR
$(CH_3)_2CHOCF_2S-$	—	7.2 μ
$-CF=S$	—	8.2 μ
$-CF=O$	—	5.4 μ

It is significant to note that polymer obtained by dimethylformamide initiation does not contain initiator fragments. Polymer obtained by initiation with tetraisopropyl titanate contains an isopropoxy end that is hydrolyzable to $-SCF=O$.

The anionic polymerization of thiocarbonyl fluoride appears to procede by the scheme given below in which $B^{\ominus}$ denotes the anion supplied by the initiator.

Initiation: $\quad B^{\ominus} + CF_2=S \longrightarrow BCF_2S^{\ominus}$

Propagation: $BCF_2-S^{\ominus} + nCF_2=S \longrightarrow B-(CF_2S)_n-CF_2S^{\ominus}$

Termination: $B-(CF_2S)_n-CF_2S^{\ominus} \longrightarrow B-(CF_2S)_n-\overset{\overset{\displaystyle S}{\|}}{CF} + F^{\ominus}$

$F^{\ominus}$ is also a base capable of initiating polymerization.

$$F^{\ominus} + CF_2=S \longrightarrow CF_3S^{\ominus}$$

$$CF_3S^{\ominus} + nCF_2=S \longrightarrow CF_3S-(CF_2S)_n-CF_2S^{\ominus}$$

$$CF_3S-(CF_2S)_n-CF_2S^{\ominus} \longrightarrow CF_3S-(CF_2S)_n-\overset{\overset{\displaystyle S}{\|}}{CF} + F^{\ominus}$$

When tetraisopropyl titanate is used as the initiator, $B^{\ominus}$ is $(CH_3)_2CHO^{\ominus}$. As indicated earlier, the polymer obtained has isopropoxy on the leading end and CF=S on the terminal end. Hydrolysis of the isopropoxy is responsible for CF=O and chain transfer as a result of $F^{\ominus}$ initiation leads to CF_3S- ends.

Dimethylformamide initiation is a special case. Polythiocarbonyl fluoride prepared with this initiator, as has been stated, also has all three possible ends with $-SCF=O$ being present in smallest amount. Also, the polymer as isolated is free of the formamide. It is here proposed that dimethylformamide initiation may proceed as follows:

$$\text{Initiation:} \quad (CH_3)_3N\overset{O}{\overset{\|}{C}}H + CF_2=S \longrightarrow (CH_3)_2\overset{\oplus}{N}=CH\overset{O-CF_2S^{\ominus}}{|}$$

$$\text{Propagation:} \quad (CH_3)_2\overset{\oplus}{N}=CH\overset{OCF_2S^{\ominus}}{|} + nCF_2=S \longrightarrow (CH_3)_2\overset{\oplus}{N}=CH\overset{O(CF_2S)_nCF_2S^{\ominus}}{|}$$

$$\text{Termination:} \quad (CH_3)_2\overset{\oplus}{N}=CH\overset{O(CF_2S)_nCF_2S^{\ominus}}{|} \longrightarrow (CH_3)_2\overset{\oplus}{N}=CH\overset{O(CF_2S)_n\overset{S}{\overset{\|}{C}}F}{|} + F^{\ominus}$$

$$\text{Transfer:} \quad F^{\ominus} + CF_2=S \longrightarrow CF_3S(CF_2S)_nCF_2\overset{\ominus}{S} \longrightarrow$$
$$CF_3S(CF_2S)_n\overset{S}{\overset{\|}{C}}F + F^{\ominus}$$

CF=O ends could be formed by hydrolysis during work-up of the first formed polymer molecules.

$$(CH_3)_2\overset{\oplus}{N}=CH\overset{O(CF_2-S)_n\overset{S}{\overset{\|}{C}}F}{|} \xrightarrow{H_2O} (CH_3)_2N\overset{O}{\overset{\|}{C}}H + HF + F\overset{O}{\overset{\|}{C}}(CF_2S)_n\overset{S}{\overset{\|}{C}}F$$

Degradation of the polymer by heat and by amines has been explained (58) as involving conversion of a thioacid fluoride end to a labile group.

$$CF_3S(CF_2S)_nCF_2\overset{S}{\overset{\|}{S}}CF \xrightarrow[H_2O]{\text{occluded}} CF_3S(CF_2S)_nCF_2S-C\overset{\nearrow O}{\underset{\searrow S}{}}\ominus$$

$$\downarrow \Delta$$

$$CF_3S(\overset{\frown}{CF_2S})_nCF_2\overset{\frown}{S^{\ominus}} + COS$$

$$\downarrow$$

$$CF_3S^{\ominus} + CF_2=S$$
$$\downarrow$$
$$CF_2=S + F^{\ominus}$$

$$CF_3S(CF_2S)_nCF_2\overset{S}{\overset{\|}{S}}-CF \xrightarrow[H_2O]{Et_3N} CF_3S(CF_2S)_nCF_2-S-C\overset{\nearrow S}{\underset{\searrow O}{}}\ominus$$
$$Et_3NH^{\oplus}$$
$$+ Et_3N \cdot HF$$

Support for this explanation can be found in the oligomers having CF_3 on both ends, which have good heat stability and do not degrade in the presence of amines, and in the fact that glass containing poly(thiocarbonyl fluoride) becomes etched upon storage because of HF formed from reaction of the polymer with water occluded on the glass.

Poly(thiocarbonyl fluoride) as isolated from the polymerization reaction mixture is amorphous. Its glass transition temperature is $-118°$ C (torsion pendulum) and its crystalline melting point is $35°$ C. It can be fabricated into films or other objects by compression molding at $100-150°$ C. Because molecular weight is so high, it is usually desirable to keep material in the mold for several hours to allow the polymer to flow well enough to be "nerve" free. These molded objects are elastomeric. Indeed, poly(thiocarbonyl fluoride) is one of the most resilient materials known. At room temperature Yerzley oscillograph measurements (ASTM D-945) have given resilience values of 95%. However, the polymer slowly crystallizes to an opaque, white, non-resilient form. It can be reconverted to the elastomeric form by reheating above $35°$ C. Because this change from an elastomer to a plastic at temperatures below $350°$ C is a disadvantage for many uses, much effort has been spent on copolymerization as a means of obtaining lower melting products. This will be discussed later.

Because of its chemical inertness, no direct way of curing poly(thiocarbonyl fluoride) has been found. However, creep has been reduced and strength at elevated temperatures improved by milling into the polymer a free-radical generator, such as dicumyl peroxide or azobisisobutyronitrile, and a free-radical acceptor, such as N,N'-*m*-phenylenebismaleimide or triacryloylhexahydro-*s*-triazine, and curing with heat and pressure (65). A better method is to mill in divinylbenzene and a small amount of benzoyl peroxide and cure with heat and pressure (66). The divinylbenzene forms a crosslinked matrix that mechanically traps poly(thiocarbonyl fluoride) molecules. Since the elastomer is in effect filled with poly(divinyl benzene), the final composition is less resilient than untreated poly(thiocarbonyl fluoride).

2. Copolymers of Thiocarbonyl Fluoride

Thiocarbonyl fluoride copolymerizes in ionic systems with fluorothioacyl fluorides, hexafluorocyclobutanone, perfluoroacetaldehyde and perfluoropropionaldehyde (59, 67). Anionic copolymerizations of chlorofluorothioacetyl fluoride using dimethylformamide are carried out in the same manner as homopolymerization of thiocarbonyl fluoride. The products are high molecular weight elastomers containing 1–3% of the thioacid fluoride. They are snappy but they are not as resilient as poly(thiocarbonyl fluoride). Crystallization temperatures of the copolymers are much lower than the $35°$ C freezing temperature of the homopolymer. Copolymers containing 3% of the thioacid fluoride crystallize below $0°$ C.

The thermal and chemical stability of the chlorofluorothioacetyl fluoride copolymers are considerably better than the homopolymer. It has been suggested that these effects are associated with ability of the side group to interrupt chain decomposition. When the chain degrades back to a thioacyl unit, the reaction stops because elimination of a fluoride ion leaves an end resistant to further degradation.

$$
\begin{array}{c}
\text{H}-\overset{\overset{\displaystyle F}{|}}{\underset{\underset{\displaystyle F}{|}}{C}}-\text{Cl} \\
\text{CF}_3\text{S}\mathord{\sim\!\sim\!\sim}\text{C}-\text{S}\mathord{-}(\text{CF}_2-\text{S})_n\text{CF}=\text{S} \xrightarrow[\text{H}_2\text{O}]{\text{occluded}}
\end{array}
$$

$$
\text{CF}_3\text{S}\mathord{\sim\!\sim\!\sim}\overset{\overset{\displaystyle H-C-Cl}{|}}{\underset{\underset{\displaystyle F}{|}}{C}}-\text{S}-(\text{CF}_2-\text{S})_n\text{C}\!\!\begin{array}{c}\diagup\!\!\!\diagup\text{O}\\ \diagdown\text{S}^{\ominus}\end{array}
$$

$$\xrightarrow{-\text{COS}}$$

$$
\text{CF}_3\text{S}\mathord{\sim\!\sim\!\sim}\overset{\overset{\displaystyle H-C-Cl}{|}}{\underset{\underset{\displaystyle F}{|}}{C}}-\text{S}-(\text{CF}_2\overset{\frown}{-\text{S}})_{n-1}\text{CF}_2\overset{\frown}{-\text{S}^{\ominus}} \xrightarrow{-\text{CF}_2=\text{S}}
$$

$$
\text{CF}_3\text{S}\mathord{\sim\!\sim\!\sim}\overset{\overset{\displaystyle H-C-Cl}{|}}{\underset{\underset{\displaystyle F}{|}}{C}}-\text{S}^{\ominus} \longrightarrow \text{CF}_3\text{S}\mathord{\sim\!\sim\!\sim}\overset{\overset{\displaystyle H-C-Cl}{|}}{\underset{\displaystyle F}{C}}=\text{S}
$$

Copolymerization of thiocarbonyl fluoride and perfluorothioacetyl fluoride, $CF_3CF=S$, have been done in ether at $-80°$ C using N-methylmorpholine and tetraisopropyl titanate as initiators. The products, which contain 13–40% of the acid fluoride, are of relatively low molecular weight, do not melt very much lower than the homopolymer, and are much less elastic than the copolymer.

Copolymers with hexafluorocyclobutanone have been made in ether at low temperatures using cesium fluoride as an initiator. They are of low molecular weight and are more resistant to amine degradation than the homopolymer. Copolymerization of perfluoroacetaldehyde and perfluoropropionaldehyde with thiocarbonyl fluoride has been accomplished by initiation with complex compounds of the type $Me(PR_3)_nB$ (67) in which Me is a group VIII metal and B is $P(C_6H_5)_3$ or a halogen. The products have excellent resistance to acids.

3. Polymerization of Fluorothioacid Fluorides and Fluorothioketones

Copolymerization of fluorothioacyl fluorides with thiocarbonyl fluoride has just been discussed. These compounds also homopolymerize. The procedure used for polymerization of thiocarbonyl fluoride, which is dimethylformamide initiation in ether at $-78°$ C, is effective for chlorofluorothioacetyl fluoride, difluorothioacetyl fluoride, trifluorothioacetyl fluoride, chlorodifluorothioacetyl fluoride, and perfluorothiopropionyl fluoride. Polymers from the first three of these monomers are elastomeric at room temperature. They have not been investigated to the same extent as polythiocarbonyl fluoride.

Chlorodifluorothioacetyl chloride has also been polymerized by the same method as described above. Films of the polymer are white, tough, and flexible, but they are not elastomeric.

Pentafluoro-3-butenethioyl fluoride polymerizes spontaneously when kept in glass for long periods. It is elastomeric and retains rubbery character when heated to its decomposition temperature, which is 240° C.

Fluorothioketones are more difficult to polymerize. There are two reasons. First, agents that promote polymerization also catalyze dimerization to dithietanes, which is a very fast reaction. Second, ceiling temperature of polymerization is low with the result that polymer decomposes back to monomer as it is being isolated. However, poly(hexafluorothioacetone) can be formed at very low temperatures by initiation with dimethylformamide or BF_3 etherate, even though at $-78°$ C the only product isolated is 2,2,4,4-tetrakis(trifluoromethyl)-1,3-dithietane.

$$\underset{CF_3}{\overset{CF_3}{\diagup}}C=S \quad \begin{array}{c} \xrightarrow{-78°C} \\ \\ \xrightarrow{-110°C} \end{array} \quad \begin{array}{c} CF_3 \diagdown \diagup S \diagdown \diagup CF_3 \\ C \quad\quad C \\ CF_3 \diagup \diagdown S \diagdown CF_3 \\ \\ \underset{CF_3}{\overset{CF_3}{|}} \\ \ \ \ (C-S)_n \\ \end{array}$$

Pressed films of the polymer are white and elastic. When kept at room temperature for several days, the polymer slowly degrades. Eventually it is all converted to the dithietane. It must be emphasized that these products should be handled with great care because the dithietane is a very toxic compound.

D. Fluorothiocarbonyl Polymers by Free-Radical Initiation

1. Thiocarbonyl Chlorofluoride

One of the great surprises of fluorothiocarbonyl chemistry is the ease with which these compounds undergo free-radical polymerization. This behavior is unique among thiocarbonyl compounds. Though thioacetone polymerizes in free-radical systems, it does not do so with anything like the avidity of fluorothiocarbonyl compounds. Thioacetone does not copolymerize with compounds containing carbon-carbon unsaturation, which is a most important property of fluorothiocarbonyl compounds.

Propensity for free-radical polymerization is most marked for thiocarbonyl chlorofluoride (39). Exposure of the monomer to a sunlamp, especially if benzoin methyl ether is present, leads to rather rapid formation of polymer. Monomer cooled with solid carbon dioxide is also converted to polymer by high-energy electrons. This high yield conversion at low temperatures shifted attention to polymerization with chemically generated radicals at low temperatures.

The main problem is how to generate free radicals at low temperatures. It was discovered this can be done by using the trialkylborane-oxygen redox couple. Prior to the studies on thiocarbonyl compounds, Furukawa and Tsuruta (68) had used a mixture of trialkylboranes and oxygen for vinyl polymerizations, and studies by Fordham and Sturm (69) and Zutty and Welch (70) had confirmed them as free-radical polymerizations. For the fluorothiocarbonyl work (39), it was shown that at $-78°$ C the reaction of a trialkylborane and oxygen proceeds cleanly to an alkylperoxydialkylborane, V.

$$R_3B + O_2 \longrightarrow R_2BOOR$$
$$V$$

If excess trialkylborane is present, a second reaction takes place that generates free radicals.

$$R_2BOOR + R'_3B \longrightarrow R_2BOBR'_2 + R'_2BOR + 2R'\cdot$$

Thus, the initiator can be a trialkylborane to which less than one molecule of oxygen per molecule of borane is added, or it can be an alkylperoxydialkylborane to which additional trialkylborane is added. In the latter case, it is convenient to use a standardized hexane solution of an alkylperoxydialkylborane, which is stable for about two weeks if stored at $-78°$ C. It is important to recognize that use of excess oxygen with the borane in the former case will result in no initiation or very poor initiation.

The trialkylborane-oxygen couple leads to rapid formation of thiocarbonyl chlorofluoride homopolymer at $-80°$ C (39). The polymer is a tough elastomer that does not crystallize when stretched at $0°$ C or when relaxed and cooled to $-80°$ C.

This monomer polymerizes too rapidly in the presence of free radicals to copolymerize easily with most vinyl monomers. However, it readily forms copolymers with 2,3-dichloro-1,3-butadiene, which also is a monomer that polymerizes very rapidly.

2. Thiocarbonyl Fluoride

The low temperature peroxyborane system is very effective for converting thiocarbonyl fluoride to homopolymer. The product is comparable to those formed by anionic polymerization. Since polymerization of thiocarbonyl fluoride is substantially slower than that of the chlorofluoride, this monomer copolymerizes with exceptional ease with a large number of vinyl compounds to give products that appear to be random copolymers.

Monomers that copolymerize with thiocarbonyl fluoride include olefins, vinyl halides, vinyl esters, allyl esters, acrylates, vinyl ethers, and vinyltrichlorosilane. Nonconjugated diolefins lead to crosslinked products. Conjugated dienes inhibit polymerization.

Propylene copolymerization is a remarkable case. Copolymers with molecular weights as high as 800000 are readily obtained. The copolymerization tends toward a product containing two molecules of thiocarbonyl fluoride for each propylene molecule. Compositions with higher thiocarbonyl content can be obtained by use of less propylene in the starting monomer mixture. Products containing approximately 2:1 thiocarbonyl fluoride/propylene ratios are soft, pliable elastomers that retain flexibility to $-55°$ C.

Ethylene, isobutylene, tert-butylethylene, and other olefins also copolymerize with thiocarbonyl fluoride. What is astonishing is that tetramethylethylene, which is so sterically hindered as to be unreactive in olefin polymerizations, copolymerizes readily with thiocarbonyl fluoride. Most of these comonomers do not behave in the same way as propylene. They give products with compositions more closely approximating the monomer mixture used.

Vinyl acetate copolymers, when hydrolyzed, undergo extensive degradation. This behavior indicates that the acetate groups and the thioether link are on the same carbon atom as in VI.

$$\begin{array}{c} \text{OAc} \\ | \\ \sim\sim CH_2CHS-CF_2 \sim\sim \\ \text{VI} \end{array} \longrightarrow \begin{array}{c} \text{OH} \\ | \\ \sim\sim CH_2CH-SCF_2 \sim\sim \end{array}$$

$$\downarrow$$

$$\sim\sim CH_2\overset{O}{\overset{\|}{C}}H + HSCF_2 \sim\sim$$

$$\downarrow {\scriptstyle -HF}$$

$$S=CF \sim\sim$$

This explanation is supported by the fact that 3-butenyl acetate copolymers do not degrade upon hydrolysis.

$$\begin{array}{c} \text{OAc} \\ | \\ CH_2 \\ | \\ CH_2 \\ | \\ \sim\sim CH_2-CH-S-CF_2 \sim\sim \end{array} \longrightarrow \begin{array}{c} \text{OH} \\ | \\ CH_2 \\ | \\ CH_2 \\ | \\ -CH_2-CH-SCF_2 \sim\sim \end{array}$$

$$\downarrow$$

no further reaction

Rationalization of these facts appear to require an abnormal addition. It has been proposed (*39*) that a free-radical adds to thiocarbonyl fluoride first to give a new radical, VII, that adds to vinyl acetate to give VIII, which then adds to the thiocarbonyl compound.

$$S{=}CF_2 \xrightarrow{R\cdot} \underset{\text{VII}}{RSCF_2\cdot} \xrightarrow{CH_2{=}CH\text{-}OAc} \underset{\text{VIII}}{RSCF_2CH_2\overset{\displaystyle OAc}{\underset{|}{C}}H\cdot}$$

$$\downarrow CF_2{=}S$$

$$RSCF_2CH_2{-}\overset{\displaystyle OAc}{\underset{|}{C}}HSCF_2\cdot$$

$$\downarrow \text{repeat}$$

$$\text{copolymer}$$

Copolymers offer the twin advantages of reduced crystalline melting temperature and introduction of crosslinking sites. One of the best for both of these purposes that has been reported so far is the thiocarbonyl fluoride-allyl chloroformate copolymer (39). Products containing as little as 2–3 mole % of allyl chloroformate melt below 0° C and are readily crosslinked by incorporation of zinc oxide followed by heating.

VI. References

1. Fukuyawa, M., Ohno, A.: Kagaku No Ryoiki **1968**, 22 (12), 1091.
2. Potts, K. T., Sapino, C.: In: Patai, S. (Ed.): The chemistry of acyl halides, Chapter 11, pp. 349—380. London: Interscience 1972.
3. Wohl, A.: Ber. **19**, 2344 (1886).
4. Le Fèbre, R. J. W., Macleod, M.: J. Chem. Soc. **1931**, 474.
5. Harmon, J.: U. S. Patent 3 070 580 (Dec. 25, 1962, Application July 16, 1959).
6. Cairns, T. L., Evans, G. L., Larchar, A. W., McKusick, B. C.: J. Am. Chem. Soc. **74**, 3982 (1952).
7. Hofmann, A. W.: Liebigs Ann. **145**, 360 (1868).
8. Credali, L., Russo, M.: Polymer **8** (9), 469 (1967).
9. Credali, L., Mortillaro, L., Russo, M., De Checchi, C.: J. Appl. Polymer Sci. **10**, 859 (1966).
10. Gipstein, E., Wellisch, E., Sweeting, O. J.: J. Polymer Sci. B **1** (5), 237 (1963).
11. Birkner, H., Stuerzenhofecker, F.: German Patent 1 202 500 (Oct. 7, 1965).
12. Stannett, V.: AEC Accession No. 37176, Rep. No. BNL-874, Arail. OTS, 2 p. (1964).
13. Lando, J. B., Stannett, V.: J. Polymer Sci. B **2**, (4), 375 (1964); A **3** (6), 2369 (1965); paper presented to American Chemical Society, Division of Polymer Chemistry, Chicago, Illinois, Sept. 1964; Polymer Preprints **5**, No. 2, 969—974 (1964).
14. Carazzolo, G., Mammi, M.: J. Polymer Sci. B **2** (11), 1057 (1964).
15. Nadkarni, V. M., Schultz, J. M.: J. Mater. Sci. **1973**, 8 (4), 525—538.
16. Schmidt, M., Blaettner, K.: Angew. Chem. **71**, 4078 (1959).
17. Russo, M., Mortillaro, L., De Checchi, C., Credali, L.: Gazz. Chim. Ital. **95**, 448 (1965).
18. Russo, M., Mortillaro, L., Credali, L., De Checchi, C.: J. Polymer Sci. B **3**, 455 (1965).
19. Lal, J.: J. Org. Chem. **26**, 971 (1961).
20. Mortillaro, L., Russo, M., Credali, L., De Checchi, C.: J. Chem. Soc. C **1966**, 428.
21. Bapseres, P., Signouret, J.: French Patent 1 330 819 (June 28, 1963).
22. Fabre, R., Verne, J., Chaigneau, M., Le Moan, G.: Compt. Rend. **259** (15), 2545 (1964).
23. Carazzolo, G., Mortillaro, L., Credali, L., Bezzi, S.: Chim. Industr. **46**, 1484 (1964).
24. Carazzolo, G., Valle, G.: Makromol. Chem. **90**, 66 (1966).
25. Carazzolo, G.: J. Polymer Sci. A **1**, 1573 (1963).
26. Bapseres, P., Audouze, B., Signouret, J., Ourgand, J.: French Patent 1 362 500 (June 5, 1964).
27. Audouze, B.: French Patent 1 383 946 (Jan. 4, 1965); Netherlands Patent 135 863 (July 17, 1972).
28. Molinet, G., Audouze, B.: French Patent 1 373 025 (Sept. 25, 1964).
29. Gourdon, B.: French Patent 1 412 409 (Oct. 1, 1965); Belgian Patent 660 559 (Sept. 3, 1965).
30. Lipscomb, R. D., Sharkey, W. H.: J. Polymer Sci. A-1, **8**, 2187 (1970).
31. Baumann, E., Fromm, E.: Ber. **22**, 2592 (1889).
32. Ettinghausen, O. G. von, Kendrick, E.: Polymer **7**, 469 (1966).
33. Burnop, V. C. E., Latham, K. G.: Polymer **8**, 589 (1967).
34. Bailey, W. J., Chu, H.: Paper presented to American Chemical Society, Division of Polymer Chemistry, Detroit, Michigan, April 1965; Polym. Preprints **6**, 145 (1965).
35. Mayer, R., Berthold, H.: Chem. Ber. **96**, 3096 (1963).
36. Mayer, R., Morganstern, J., Fabian, J.: Angew. Chem. Intern. Ed. **3**, 277 (1964).
37. Roo, C. N. R., Venkataraghavan, R.: Spectrochim. Acta **18**, 541 (1962).
38. Haszeldine, R. N., Kidd, J. M.: J. Chem. Soc. **1955**, 3871.
39. Barney, A. L., Bruce, J. M. Jr., Coker, J. N., Jacobson, H. W., Sharkey, W. H.: J. Polymer Sci. A-1, **4**, 2617 (1966).
40. Ettinghausen, O. G. von: French Patent 1 425 651 (Jan. 21, 1966) (Esso Research and Engineering Company); Brit. Appl. Feb. 14, 1964, and Feb. 5, 1965.
41. Kunitake, T., Yasumatsu, M., Aso, C.: J. Polymer Sci. A-1, **9**, 3675 (1971).
42. Iwakura, Y., Nabeya, A., Nishiguchi, T.: J. Polymer Sci. A-1, **6**, 2591 (1968).
43. Sharkey, W. H.: Carbonyl and thiocarbonyl compounds. In: Stille, J. K., Campbell, T. W. (Eds.): Condensation monomers, high polymer, Vol. 27, p. 651. New York: John Wiley and Sons, Inc. 1972.

44. Middleton, W. J., Howard, E. G., Sharkey, W. H.: J. Org. Chem. **30**, 1375 (1965).
45. Sharkey, W. H., Jacobson, H. W.: In: Wittbecker, E. L. (Ed.): Macromolecular syntheses, Vol. 5, p. 25. New York: John Wiley and Sons, Inc. 1974.
46. Downs, A. J.: Spectrochim. Acta **19**, 1165 (1963).
47. Hopper, M. J., Russell, J. W., Ovesend, J.: Spectrochim. Acta **28** A, 1215 (1972).
48. Kroto, H. W., Suffolk, R. J.: Chem. Physics Letters **17** (2), 213 (1972).
49. Careless, A. J., Kroto, H. W., Landsberg, B. M.: Chem. Phys. **1** (4), 371 (1973).
50. Yarovenko, N. N., Vasil'eva, A. S.: J. Gen. Chem. (USSR) **29**, 3754 (1959) (english translation).
51. Downs, A. J., Ebsworth, E. A. V.: J. Chem. Soc. **1960**, 3516.
52. Marquis, D. M.: U.S. Patent 2962529 (Sept. 29, 1960).
53. Martin, K. V.: U.S. Patent 3048629 (1962).
54. Yarovenko, N. N., Motornyi, S. P., Kirenskaya, L. I., Vasilyeva, A. S.: J. Gen. Chem. (USSR) **27**, 2301 (1957) (english translation).
55. Harris, J. F. Jr., Stacey, F. W.: J. Am. Chem. Soc. **85**, 749 (1963).
56. Sharkey, W. H.: In: Stille, J. K. (Ed.): Condensation monomers, p. 684. New York: John Wiley and Sons, Inc. 1972.
57. Dyatkin, B. L., Sterlin, S. R., Zhuravkova, L. G., Martynov, B. I., Mysov, E. I., Knunyants, I. L.: Tetrahedron **1973**, 29 (18), 2759—2767.
58. Middleton, W. J., Jacobson, H. W., Putnam, R. E., Walter, H. C., Acker, D. S., Sharkey, W. H.: Abstracts of Papers, 145th Meeting of the American Chemical Society, September, 1963, New York, N.Y.
59. Middleton, W. J., Jacobson, H. W., Putnam, R. E., Walter, H. C., Pye, D. G., Sharkey, W. H.: J. Polymer Sci. A **3**, 4115 (1965).
60. Sharkey, W. H.: Am. Chem. Soc., Div. Polymer Chem. Preprints **7** (1), 275—278 (1966); J. Macromol. Sci. (Chem.) A **1**, (2), 291 (1967).
61. Gubanov, V. A., Varsharskii, Y. S., Dolgopol'skii, I. M., Brettske, E. B.: U.S.S.R. Patent 265442 (March 9, 1970); C.A. **73**, 12 (1970), No. 26037 z.
62. Ya Poddubnyi, I., Dolgopol'skii, I. M., Evdokimov, V. F., Gubanov, V. A., Tumanova, A. V., Brettske, E. G.: U.S.S.R. Patent 241010 (Sept. 10, 1969); C.A. **72**, 18 (1970); No. 44348 k.
63. Sharkey, W. H.: Chem. Sulfides Conf. **1966** (Pub. 1968), 205—210, Edited by Arthur V. Tobolsky. New York, N.Y.: Interscience Publishers.
64. Sharkey, W. H.: In: Kennedy, J. P., Tornqvist, E. G. M. (Eds.): Polymer chemistry of synthetic elastomers, Part 2, p. 893. New York, N.Y.: Interscience Publishers.
65. Acker, D. S.: U.S. Patent 3297790 (Jan. 10, 1967); C.A. **67**, 1195 (1967), No. 12372 u.
66. Acker, D. S., Barney, A. L.: U.S. Patent 3378604 (April 16, 1968); C.A. **68**, 11150 (1968), No. 115548 f.
67. Gubanov, V. A., Dolgopol'skii, I. M., Brettski, E. B.: U.S.S.R. Patent 304270 (July 15, 1971).
68. Furukawa, J., Tsuruta, T.: J. Polymer Sci. **28**, 227 (1958).
69. Fordham, J. W. L., Sturm, C. S.: J. Polymer Sci. **33**, 503 (1958).
70. Zutty, N. K., Welch, F. J.: J. Polymer Sci. **43**, 445 (1960).

Received October 2, 1974

Author-Index Volume 1—17

ALLEGRA, G. and BASSI, I. W.: Isomorphism in Synthetic Macromolecular Systems. Vol. 6, pp. 549—574.
AYREY, G.: The Use of Isotopes in Polymer Analysis. Vol. 6, pp. 128—148.
BALDWIN, R. L.: Sedimentation of High Polymers. Vol. 1, pp. 451—511.
BERGSMA, F. and KRUISSINK, CH. A.: Ion-Exchange Membranes. Vol. 2, pp. 307—362.
BERRY, G. C. and FOX, T. G.: The Viscosity of Polymers and their Concentrated Solutions. Vol. 5, pp. 261—357.
BEVINGTON, J. C.: Isotopic Methods in Polymer Chemistry. Vol. 2, pp. 1—17.
BIRD, R. B., WARNER, JR., H. R., and EVANS, D. C.: Kinetic Theory and Rheology of Dumbbell Suspensions with Brownian Motion. Vol. 8, pp. 1—90.
BÖHM, L. L., CHMELIŘ, M., LÖHR, G., SCHMITT, B. J. und SCHULZ, G. V.: Zustände und Reaktionen des Carbanions bei der anionischen Polymerisation des Styrols. Vol. 9, pp. 1—45.
BOVEY, F. A. and TIERS, G. V. D.: The High Resolution Nuclear Magnetic Resonance Spectroscopy of Polymers. Vol. 3, pp. 139—195.
BREITENBACH, J. W., OLAJ, O. F. und SOMMER, F.: Polymerisationsanregung durch Elektrolyse. Vol. 9, pp. 47—227.
BRESLER, S. E. and KAZBEKOV, E. N.: Macroradical Reactivity Studied by Electron Spin Resonance. Vol. 3, pp. 688—711.
BYWATER, S.: Polymerization Initiated by Lithium and its Compounds. Vol. 4, pp. 66—110.
CARRICK, W. L.: The Mechanism of Olefin Polymerization by Ziegler-Natta Catalysts. Vol. 12, pp. 65—86.
CASALE, A. and PORTER, R. S.: Mechanical Synthesis of Block and Graft Copolymers. Vol. 17, pp. 1—71.
CERF, R.: La dynamique des solutions de macromolécules dans un champ de vitesses. Vol. 1, pp. 382—450.
CICCHETTI, O.: Mechanisms of Oxidative Photodegradation and of UV Stabilization of Polyolefins. Vol. 7, pp. 70—112.
COLEMAN, JR., L. E. and MEINHARDT, N. A.: Polymerization Reactions of Vinyl Ketones. Vol. 1, pp. 159—179.
CRESCENZI, V.: Some Recent Studies of Polyelectrolyte Solutions. Vol. 5, pp. 358—386.
DOLE, M.: Calorimetric Studies of States and Transitions in Solid High Polymers. Vol. 2, pp. 221—274.
DUŠEK, K. and PRINS, W.: Structure and Elasticity of Non-Crystalline Polymer Networks. Vol. 6, pp. 1—102.
EASTHAM, A. M.: Some Aspects of the Polymerization of Cyclic Ethers. Vol. 2, pp. 18—50.
EHRLICH, P. and MORTIMER, G. A.: Fundamentals of the Free-Radical Polymerization of Ethylene. Vol. 7, pp. 386—448.
EISENBERG, A.: Ionic Forces in Polymers. Vol. 5, pp. 59—112.
ELIAS, H.-G., BAREISS, R. und WATTERSON, J. G.: Mittelwerte des Molekulargewichtes und anderer Eigenschaften. Vol. 11, pp. 111—204.
FISCHER, H.: Freie Radikale während der Polymerisation, nachgewiesen und identifiziert durch Elektronenspinresonanz. Vol. 5, pp. 463—530.
FUJITA, H.: Diffusion in Polymer-Diluent Systems. Vol. 3, pp. 1—47.
FUNKE, W.: Über die Strukturaufklärung vernetzter Makromoleküle, insbesondere vernetzter Polyesterharze, mit chemischen Methoden. Vol. 4, pp. 157—235.
GAL'BRAIKH, L. S. and ROGOVIN, Z. A.: Chemical Transformations of Cellulose. Vol. 14, pp. 87—130.
GERRENS, H.: Kinetik der Emulsionspolymerisation. Vol. 1, pp. 234—328.

GRAESSLEY, W. W.: The Etanglement Concept in Polymer Rheology. Vol. 16, pp. 1—179.
HAY, A. S.: Aromatic Polyethers. Vol. 4, pp. 496—527.
HAYAKAWA, R. and WADA, Y.: Piezoelectricity and Related Properties of Polymer Films. Vol. 11, pp. 1—55.
HELFFERICH, F.: Ionenaustausch. Vol. 1, pp. 329—381.
HENDRA, P. J.: Laser-Raman Spectra of Polymers. Vol. 6, pp. 151—169.
HENRICI-OLIVÉ, G. und OLIVÉ, S.: Kettenübertragung bei der radikalischen Polymerisation. Vol. 2, pp. 496—577.
HENRICI-OLIVÉ, G. und OLIVÉ, S.: Koordinative Polymerisation an löslichen Übergangs-metall-Katalysatoren. Vol. 6, pp. 421—472.
HERMANS, JR., J., LOHR, D., and FERRO, D.: Treatment of the Folding and Unfolding of Protein Molecules in Solution According to a Lattic Model. Vol. 9, pp. 229—283.
HUTCHISON, J. and LEDWITH, A.: Photoinitiation of Vinyl Polymerization by Aromatic Carbonyl Compounds. Vol. 14, pp. 49—86.
ISE, N.: Polymerizations under an Electric Field. Vol. 6, pp. 347—376.
ISE, N.: The Mean Activity Coefficient of Polyelectrolytes in Aqueous Solutions and Its Related Properties. Vol. 7, pp. 536—593.
ISIHARA, A.: Intramolecular Statistics of a Flexible Chain Molecule. Vol. 7, pp. 449—476.
ISIHARA, A.: Irreversible Processes in Solutions of Chain Polymers. Vol. 5, pp. 531—567.
ISIHARA, A. and GUTH, E.: Theory of Dilute Macromolecular Solutions. Vol. 5, pp. 233—260.
JANESCHITZ-KRIEGL, H.: Flow Birefringence of Elastico-Viscous Polymer Systems. Vol. 6, pp. 170—318.
KENNEDY, J. P. and GILLHAM, J. K.: Cationic Polymerization of Olefins with Alkylaluminum Initators. Vol. 10, pp. 1—33.
KENNEDY, J. P. and LANGER, JR., A. W.: Recent Advances in Cationic Polymerization. Vol. 3, pp. 508—580.
KENNEDY, J. P. and OTSU, T.: Polymerization with Isomerization of Monomer Preceding Propagation. Vol. 7, pp. 369—385.
KENNEDY, J. P. and RENGACHARY, S.: Correlation between Cationic Model and Polymerization Reactions of Olefins. Vol. 14, pp. 1—48.
KITAGAWA, T. and MIYAZAWA, T.: Neutron Scattering and Normal Vibrations of Polymers. Vol. 9, pp. 335—414.
KNAPPE, W.: Wärmeleitung in Polymeren. Vol. 7, pp. 477—535.
KONINGSVELD, R.: Preparative and Analytical Aspects of Polymer Fractionation. Vol. 7, pp. 1—69.
KOVACS, A. J.: Transition vitreuse dans les polymères amorphes. Etude phénoménologique. Vol. 3, pp. 394—507.
KRÄSSIG, H. A.: Graft Co-Polymerization to Cellulose and its Derivatives. Vol. 4, pp. 111—156.
KRAUS, G.: Reinforcement of Elastomers by Carbon Black. Vol. 8, pp. 155—237.
KRIMM, S.: Infrared Spectra of High Polymers. Vol. 2, pp. 51—172.
KUHN, W., RAMEL, A., WALTERS, D. H., EBNER, G., and KUHN, H. J.: The Production of Mechanical Energy from Different Forms of Chemical Energy with Homogeneous and Cross-Striated High Polymer Systems. Vol. 1, pp. 540—592.
KURATA, M. and STOCKMAYER, W. H.: Intrinsic Viscosities and Unperturbed Dimensions of Long Chain Molecules. Vol. 3, pp. 196—312.
MEYERHOFF, G.: Die viscosimetrische Molekulargewichtsbestimmung von Polymeren. Vol. 3, pp. 59—105.
MORAWETZ, H.: Specific Ion Binding by Polyelectrolytes. Vol. 1, pp. 1—34.
MULVANEY, J. E., OVERBERGER, C. G., and SCHILLER, A. M.: Anionic Polymerization. Vol. 3, pp. 106—138.
OSAKI, K.: Viscoelastic Properties of Dilute Polymer Solutions. Vol. 12, pp. 1—64.
OSTER, G. and NISHIJIMA, Y.: Fluorescence Methods in Polymer Science. Vol. 3, pp. 313—331.
OVERBERGER, C. G. and MOORE, J. A.: Ladder Polymers. Vol. 7, pp. 113—150.
PATAT, F., KILLMANN, E. und SCHLIEBENER, C.: Die Adsorption von Makromolekülen aus Lösung. Vol. 3, pp. 332—393.

PETICOLAS, W. L.: Inelastic Laser Light Scattering from Biological and Synthetic Polymers. Vol. 9, pp. 285—333.
PINO, P.: Optically Active Addition Polymers. Vol. 4, pp. 393—456.
PLESCH, P. H.: The Propagation Rate-Constants in Cationic Polymerisations. Vol. 8, pp. 137—154.
POROD, G.: Anwendung und Ergebnisse der Röntgenkleinwinkelstreuung in festen Hochpolymeren. Vol. 2, pp. 363—400.
POSTELNEK, W., COLEMAN, L. E., and LOVELACE, A. M.: Fluorine-Containing Polymers. I. Fluorinated Vinyl Polymers with Functional Groups, Condensation Polymers, and Styrene Polymers. Vol. 1, pp. 75—113.
ROHA, M.: Ionic Factors in Steric Control. Vol. 4, pp. 353—392.
ROHA, M.: The Chemistry of Coordinate Polymerization of Dienes. Vol. 1, pp. 512—539.
SAFFORD, G. J. and NAUMANN, A. W.: Low Frequency Motions in Polymers as Measured by Neutron Inelastic Scattering. Vol. 5, pp. 1—27.
SCHUERCH, C.: The Chemical Synthesis and Properties of Polysaccharides of Biomedical Interest. Vol. 10, pp. 173—194.
SCHULZ, R. C. und KAISER, E.: Synthese und Eigenschaften von optisch aktiven Polymeren. Vol. 4, pp. 236—315.
SEANOR, D. A.: Charge Transfer in Polymers. Vol. 4, pp. 317—352.
SEIDL, J., MALINSKÝ, J., DUŠEK, K. und HEITZ, W.: Makroporöse Styrol-Divinylbenzol-Copolymere und ihre Verwendung in der Chromatographie und zur Darstellung von Ionenaustauschern. Vol. 5, pp. 113—213.
SEMJONOW, V.: Schmelzviscositäten hochpolymerer Stoffe. Vol. 5, pp. 387—450.
SHARKEY, W. H.: Polymerization through the Carbon-Sulfur Double Bond. Vol. 17, pp. 73—103.
SLICHTER, W. P.: The Study of High Polymers by Nuclear Magnetic Resonance. Vol. 1, pp. 35—74.
SMETS, G.: Block and Graft Copolymers. Vol. 2, pp. 173—220.
SOTOBAYASHI, H. und SPRINGER, J.: Oligomere in verdünnten Lösungen. Vol. 6, pp. 473—548.
SPERATI, C. A. and STARKWEATHER, JR., H. W.: Fluorine-Containing Polymers. II. Polytetrafluoroethylene. Vol. 2, pp. 465—495.
SPRUNG, M. M.: Recent Progress in Silicone Chemistry. I. Hydrolysis of Reactive Silane Intermediates. Vol. 2, pp. 442—464.
STILLE, J. K.: Diels-Alder Polymerization. Vol. 3, pp. 48—58.
SZWARC, M.: Termination of Anionic Polymerization. Vol. 2, pp. 275—306.
SZWARC, M.: The Kinetics and Mechanism of N-carboxy-α-amino-acid Anhydride (NCA) Polymerisation to Poly-amino Acids. Vol. 4, pp. 1—65.
SZWARC, M.: Thermodynamics of Polymerization with Special Emphasis on Living Polymers. Vol. 4, pp. 457—495.
TANI, H.: Stereospecific Polymerization of Aldehydes and Epoxides. Vol. 11, pp. 57—110.
TATE, B. E.: Polymerization of Itaconic Acid and Derivatives. Vol. 5, pp. 214—232.
TAZUKE, S.: Photosensitized Charge Transfer Polymerization. Vol. 6, pp. 321—346.
THOMAS, W. M.: Mechanism of Acrylonitrile Polymerization. Vol. 2, pp. 401—441.
TOBOLSKY, A. V. and DUPRÉ, D. B.: Macromolecular Relaxation in the Damped Torsional Oscillator and Statistical Segment Models. Vol. 6, pp. 103—127.
TOSI, C. and CIAMPELLI, F.: Applications of Infrared Spectroscopy to Ethylene-Propylene Copolymers. Vol. 12, pp. 87—130.
TOSI, C.: Sequence Distribution in Copolymers: Numerical Tables. Vol. 5, pp. 451—462.
TSUJI, K.: ESR Study of Photodegradation of Polymers. Vol. 12, pp. 131—190.
VALVASSORI, A. and SARTORI, G.: Present Status of the Multicomponent Copolymerization Theory. Vol. 5, pp. 28—58.
VOORN, M. J.: Phase Separation in Polymer Solutions. Vol. 1, pp. 192—233.
WERBER, F. X.: Polymerization of Olefins on Supported Catalysts. Vol. 1, pp. 180—191.
WICHTERLE, O., ŠEBENDA, J., and KRÁLÍČEK, J.: The Anionic Polymerization of Caprolactam. Vol. 2, pp. 578—595.
WILKES, G. L.: The Measurement of Molecular Orientation in Polymeric Solids. Vol. 8, pp. 91—136.

Wöhrle, D.: Polymere aus Nitrilen. Vol. 10, pp. 35—107.
Wolf, B. A.: Zur Thermodynamik der enthalpisch und der entropisch bedingten Entmischung von Polymerlösungen. Vol. 10, pp. 109—171.
Woodward, A. E. and Sauer, J. A.: The Dynamic Mechanical Properties of High Polymers at Low Temperatures. Vol. 1, pp. 114—158.
Wunderlich, B. and Baur, H.: Heat Capacities of Linear High Polymers. Vol. 7, pp. 151—368.
Wunderlich, B.: Crystallization During Polymerization. Vol. 5, pp. 568—619.
Wrasidlo, W.: Thermal Analysis of Polymers. Vol. 13, pp. 1—99.
Yamazaki, N.: Electrolytically Initiated Polymerization. Vol. 6, pp. 377—400.
Yoshida, H. and Hayashi, K.: Initiation Process of Radiation-induced Ionic Polymerization as Studied by Electron Spin Resonance. Vol. 6, pp. 401—420.
Zachmann, H. G.: Das Kristallisations- und Schmelzverhalten hochpolymerer Stoffe. Vol. 3, pp. 581—687.

Die Makromolekulare Chemie

An International Journal
of Macromolecular Chemistry and Physics

Founded by Hermann Staudinger, Nobel Prize Laureate in Chemistry
Editor: Werner Kern

Die Makromolekulare Chemie

- carries papers from authors working at all major universities and research centers throughout the world;
- publishes only original papers that meet high standards of excellence and significance;
- is clearly divided into: "Chemistry of Macromolecules", "Physical Chemistry of Macromolecules", and "Physics of Macromolecules";
- provides for speedier publication of important news as "Short Communications";
- gives "Summeries" in English as well as in the original language (65% of the papers being written in English, 25% in German and 10% in French).

Die Makromolekulare Chemie is published in 12 issues a year totalling about 3600 pages. Rate for 1974 (volume 175): sfr. 1260.— / DM 1056.—.
Back issues and complete-back sets are available. Please inquire.

Hüthig & Wepf Verlag,   Eisengasse 5,   CH-4001 Basel

B. Volmert

POLYMER CHEMISTRY

Translated from the German by E. H. Immergut, New York
With 630 figures. XVII, 652 pages. 1973
Cloth DM 72,—; US $29.60
ISBN 3-540-05631-9
Prices are subject to change without notice

This book gives a comprehensive coverage of the synthesis of polymers and their reactions, structure, and properties. The treatment of the reactions used in the preparation of macromolecules and in their transformation into cross-linked materials is particularly detailed and complete. The book also gives an up-to-date presentation of other important topics, such as enzymatic and protein synthesis, solution properties of macromolecules, polymer crystallization, and properties of polymers in the solid state.
The content and presentation of Professor Vollmert's book is more encompassing than most existing treatises, and its numerous figures and tables convey a wealth of data, never, however, at the expense of intellectual clarity or educational value.
The presentation is mainly on a fundamental and general level and yet the reader—student or professional—is gradually and almost casually introduced to all important natural and synthetic polymers. Complicated phenomena are explained with the aid of the simplest available examples and models in order to ensure complete understanding. However, the reader is also encouraged to think for himself and even to criticize the author's point of view.
All of the chapters have been revised and enlarged from the German edition, and many of the sections are entirely new.

Contents
Introduction. — Structural Principles. — Synthesis and Reactions of Macromolecular Compounds. — The Properties of the Individual Macromolecule. — States of Macromolecular Aggregation.

**Springer-Verlag
Berlin Heidelberg New York**

STRY LIBRARY
debrand H   53

J